*For my mother*

# Contents

# Acknowledgments

MANY PEOPLE HELPED ME IN THE RESEARCHING AND WRITING OF THE book, in particular, Thony Christie; Thomas F. Glick, emeritus professor of history at Boston University; Owen Gingerich, emeritus professor of astronomy and the history of science at Harvard University; Adam Apt, PhD; Colleen Farrell, PhD; Toby Huff, PhD. I owe a special thank-you to Ryan Birmingham for providing all the original illustrations for the book, as well as to the staff of the Harry Elkins Widener Memorial Library at Harvard, who allowed me to keep several piles of books out of circulation for several months; to the editorial team at Prometheus—Jake Bonar, Jessica McCleary, Erin McGarvey; to Hilary Hinzmann for his editorial support, and to my agent Susan Schulman. Finally and most importantly, to my wife and daughters for their constant support.

Any errors or omissions in the book are entirely the fault of the author.

# An Ancient Inheritance

*I was the first that yoked unmanaged beasts,*
*To serve as slaves with collar and with pack,*
*And take upon themselves, to man's relief,*
*The heaviest labor of his hands: and*
*Tamed to the rein and drove in wheeled cars*
*The horse, of sumptuous pride the ornament.*
*And those sea-wanderers with the wings of cloth,*
*The shipman's wagons, none but I contrived.*
—AESCHYLUS, *PROMETHEUS BOUND* (LINES 462–69)

"WHILE CIVILIZATIONS ARE MORTAL," HISTORIAN MAURICE DAUMAS ONCE wrote, "each one has, before succumbing to its fate, prepared a heritage of which its successors were never unaware."[1]

The successors to the lands once dominated by the Roman Empire, especially those in western Europe, certainly became aware of Rome's heritage. But for a long time during the early centuries of what are now known as the Middle Ages, it must have seemed more like they were haunted.

Consider the collapse of Rome from the smallest scale, say that of the farmer or local merchant or his slaves going to market. About fifteen miles west of Rome, you can still find the ancient harbor city of Ostia, called Ostia Antica on the tourist maps. It's no longer a port, the Mediterranean having receded a few miles since the long centuries when it was the gateway for merchants heading up the Tiber to Rome. (Though who can say if it will not soon return, given the rising sea levels due to climate change.)

You can walk among the deserted avenues and see the remains of shops where artisans sold their goods, wine merchants sold their vintages, and people living above in multistory apartments could look down at those mingling and socializing. You can sit on the worn stone steps of the amphitheater where acting troupes performed the plays of Seneca or other Roman playwrights. On either side of the avenues, some of the roofless patios still retain walls with the cracked remains of elegant marble decorated with frescoes overlooking mosaic floors made of exquisite stones imported from the furthest reaches of the empire. But Ostia Antica has been deserted since the ninth century CE, when the city was abandoned by its last inhabitants, weary of repeated invasions by pirates.

Just how quickly the civilization of Rome itself disappeared is debated among historians, but most recently Brian Ward-Perkins offered a stark example of the effects of its collapse based on the difference and quality of goods available in Britain during the few short decades between the end of the fourth and beginning of the fifth centuries CE after Roman troops withdrew from the British Isles.

> *For instance, in Britain—as ever an extreme case—every one of the building crafts introduced by the Romans, the mundane as well as the luxury ones, disappeared completely during the fifth century. There is no evidence whatsoever of the continued quarrying of building stone, nor of the preparation of mortar, nor of the manufacture and use of bricks and tiles. All new buildings in the fifth and sixth centuries, whether in Anglo-Saxon or unconquered British areas, were either of wood and other perishable materials or of drystone walling, and all were roofed in wood or thatch.*[2]

Although Britain was an extreme case, the conditions on the continent also deteriorated to the point that by the end of the sixth century, currency had disappeared and farmers in Gaul returned to bartering food and livestock and did their best to weather the onslaughts of Germanic invaders: the Goths, Visigoths, Vandals, Lombards. In the south of what is now France, the Roman city of Arles shrunk to within the diameter of its own amphitheater. Only in the eastern ends of the empire and the Middle East did civilization continue on, at a relatively slower pace of economic decline, until the invasions of Arab armies following the rise of Islam at the end of the seventh century.

Yet with all the disruptions between the fourth and sixth centuries in the West—in particular, the sacking and overrunning of the cities, many of which were ultimately abandoned and devolved into ghost towns—the poorest people carried on in rural areas. And archaeologists have found plenty of evidence that the tools of antiquity survived long enough to form the foundation of industry for the next era, which historians dubbed the early Middle Ages.

Farming tools were the most important of these for immediate survival. For the base of their subsistence diet, farmers needed to grow wheat and barley to feed themselves and what little livestock they could support. The grains they would grind, making flour for bread and for porridge—the main staple of the medieval peasant's diet well into the thirteenth century.

Grinding grain had been done by hand for centuries. The hand quern, or rotary hand mill—by no means the first tool for grinding grain but one of the most efficient—probably originated during the sixth century BCE. It consisted of two thick, disc-shaped stones, often cut from sandstone or limestone. One served as an immobile base, or bed stone, which was smooth. The upper stone, or runner, was carved on the bottom with ridges or grooves into which the grains collected as they were fed into the mill via an opening at the center of the upper stone. Rotating the upper stone with a lever by hand ground the grains to a fine degree and shed them from all sides through the grooves, where they could be collected. Such a hand-operated mill might be two feet in diameter. With larger millstones that served a wider community, the person (or animal) would drive it by walking the top stone around the lower on foot by means of a horizontal lever.

The hand quern must have been universal in the ancient world of the Mediterranean among both rich and poor. Even those fashioned for domestic use must have been formidable, requiring more than one person

3

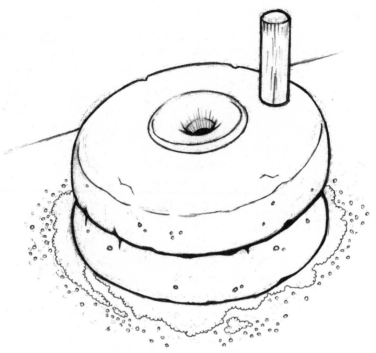

Rotary hand mill. ILLUSTRATION BY RYAN BIRMINGHAM

to move. They were certainly familiar enough to be a common reference—as we see in the Bible, the Gospel of Mark 9:42, when Jesus warned that it would be better for anyone guilty of leading a child astray to be cast in the sea with a millstone around his neck. This would have presented an immediate and vivid image to anyone hearing this at the time.

But evidence from both written records and archaeological findings suggests that the first water-powered mills were also invented not long after the hand mills—and they were built for the purpose of grinding grain on a much larger scale. This could have been as early as the third century BCE in Alexandria.[3] By the early centuries of the Common Era, the Romans constructed large networks of water-powered mills in order to feed the thousands of soldiers in their armies throughout the empire. I return to this development in the next chapter, but by the fifth century CE, this large-scale technology had disappeared along with the soldiers, craftsmen, and leaders of the empire that supported it. If any struggling farmers lived near the abandoned water mills, they may have been able to continue their operation. But most people relied on those stable tools and sources of power that were easiest to manage: their hand querns, their livestock, and plows.

It may seem odd to begin a history of medieval technology at such a basic level, with the simplest agricultural tools invented in antiquity. Compared to the technology and power sources that underlie modern civilization, developments like the heavy plow and the breast harness for horses, in conjunction with ideas put into practice such as the system of three-field crop rotation, hardly seem to merit the term "technology," let alone tools that constituted a "breakthrough." And yet, had it not been for the efficiency of these seemingly simple upgrades to already ancient farming techniques, it's unlikely the post-Roman twilight, once categorized as the Dark Ages but which now constitutes the early Middle Ages, would have ended as soon as it did. Between the ninth and thirteenth centuries, the area of lands throughout western and northern Europe available for tilling with the heavy plow expanded once farmers could successfully turn over the thickest, wettest soils. Crop yields grew with the added help of an expanding period of mild climate. And with more food and feed produced, the population began to rebound, and with it, the demand for more improvements.

But this was not something that happened overnight. Between the death of the last emperor Romulus Augustulus in 476 CE and the dawn of the Carolingian Empire in 768 CE, the struggling peasant farmers of the territories of Gaul, Germany, Britain, and northern Europe must have regarded life as a success if they were able to survive the waves of invasion, plague, and famine and still scratch out a living from the soil while one after another generation of self-appointed kings attempted to establish order.

By the time of Charlemagne, the population had rebounded to the point where the monastic orders expanded their estates and employed more farmers and craftsmen to meet the need for more food and thus the need for more efficient farming techniques and better farming tools. Indeed, the monasteries served as a bridge, adopting functions once served by cities, where there were no towns, providing markets, administration, local justice to some extent, transportation, and education, however limited in the early Middle Ages.

This need for technological advancement appears to fly in the face of a common misconception in the history of technology, especially easy to accept today, when our society seems to respond almost monthly to a new smartphone, new model of automobile, or biotechnology. The notion almost goes without saying: technology changes society. In fact, historically, it has more often been the reverse, as historian Mark Kurlansky has written. Society develops technology to address the changes that are *already* taking place within it.[4]

For purposes of this book and following the lead of other historians, I define technology in terms of "tools and techniques," which includes instruments, materials, and machines, as well as the practical skills and knowledge needed to invent, manufacture, and improve them.[5] The drive to improve—along with the inspiration to exploit natural energy sources—would follow on the increasing success of the farmers plowing the soils of western and northern Europe in the eighth and ninth centuries, and in time this would lead to a complete transformation of society in the Middle Ages.

So the ground is a good place to begin. And I should note that between the seventh and ninth centuries, there were fewer people to tend it than in antiquity. Indeed, as late as the ninth and tenth centuries, according to Georges Duby, entire countries like England and all of the Germanic territories in central Europe had no towns. "Elsewhere some towns existed: such as the few ancient Roman cities in the south which had not suffered complete dilapidation, or the new townships on trade routes which were making their appearance along the rivers leading to the northern seas."[6] This is not surprising given the past few centuries of upheaval and the vulnerability of most common people to the depredations of invading forces. According to one estimate, the population of Europe went from a high of sixty-seven million people at the peak of the Roman Empire (200 CE) to just twenty-seven million by the year 700 CE.[7] The rise of the monasteries proved to be an important bridge between the collapse of the Roman cities and the emergence of new cities in the Middle Ages.

Throughout antiquity, farmers of the Mediterranean relied on oxen to pull a simple wooden Roman plow through fields of relatively light and dry soil. Oxen were slow but sturdy and inexpensive to feed compared to horses. It was not difficult for them to drag the plow, which was often no more than a thick, whittled digging stick, known as a scratch plow. Because such plows do not turn over the soil, leaving a border of undisturbed earth between each furrow, fields need to be cross-plowed to ensure the earth is turned up enough for sowing seeds. In the warmer, drier climate of the Middle East, however, this did not present too great an inconvenience; soil was light and easy to handle for one man and one ox. And so it remained for centuries.

Such an approach would never be adequate for the struggling farmers of Gaul, Britain, and Germany, where most of the arable land consisted of dense alluvium due to the more frequent summer rains, most of which lay shrouded under forests of trees that needed to be cleared first. This was most often done by controlled burning, a happy side effect of which proved to be the ashes that helped nourish the topsoil of the newly cleared fields.

Nevertheless, it seems the first heavy plows were developed in antiquity and then later adapted by medieval Europeans. The first suggestive reference to the emergence of heavy plows comes from Pliny the Elder. Writing in the mid-first century CE, the Roman philosopher alluded to plows being drawn by teams of eight oxen in the Po Valley in the north of Italy, where, just as in the lands north of the Alps, the soil was heavier.

We don't have any more specifics about Pliny's plow, which may have been simple in design, but by the time it emerged in Europe the heavy plow had three key components. The first was the coulter, a long blade that is attached to the front of the plow pole and cuts into the earth like a vertical knife. Behind it was the plowshare, the spade-shaped edge that often formed the horizontal base of the third and final piece, the mouldboard. The plowshare and mouldboard worked in combination to turn over the soil cut by the coulter, depositing it on either side, in the same way that the bow of a boat parts the waves on either side as it moves through the water.

Drawn by one, but more often a team of two or more oxen (if some of the depictions in medieval art can be trusted in their commitment to

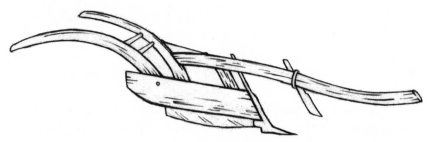

Heavy plow with coulter, plowshare, and mouldboard. ILLUSTRATION BY RYAN BIRMINGHAM

realism), a farmer with a sturdy team could plow a moderately sized field in the course of a single day. By adding a set of wheels, or carriage, in front of the coulter, the farmer could also more quickly adjust his plowing to changes in terrain.

What did farmers of the early Middle Ages plant? Medieval "corn" as it was known, was actually wheat, not to be confused with the corn people today recognize as native to the Americas. They also planted barley, rye, and oats, all of which could be used to make breads; oats became more important in the later Middle Ages as feed for horses, when they were adopted to haul plows. Peas and beans were also grown.

These general crops were markedly different from those of the southern farms of the Mediterranean and Middle East, where olive trees, fruits, and a wider variety of vegetables were the staples of the empire. "We may recall the disgust of the Gallo-Roman, Sidonius Apollinaris," Duby writes, "at the manner in which the barbarians whom he had to endure as neighbors used to eat food cooked in butter and onions. These cultures also represented two ways of exploiting natural resources, and two types of landscape coming into contact during the seventh and eighth centuries. The one was Roman and in process of decay, the other Germanic and in process of improvement, while each was gradually merging with the other."[8]

With the new plow turning over the heavy soil so thoroughly in one pass, the medieval farmer was freed from at least one extra task: cross-plowing his field as his predecessors in the south had. This saved time and energy and also allowed him to furrow longer strips.

Indeed, in his book, *Medieval Technology and Social Change*, Lynn White points out that by eliminating the need for cross-plowing, the heavy plow gradually changed the shape of the northern European landscape. Whereas crop fields in the warmer climate of the Middle East and Mediterranean were more strictly square due to cross-plowing, all across medieval Europe they were long and narrow, "with a slightly rounded vertical cross-section for each strip-field which had salutary effects on drainage in that moist climate. These strips were normally plowed clockwise, with the sod turning over and inward to the right. As a result, with the passage of the years, each strip became a long low ridge, assuring a crop on the crest even in the wettest years, and in the intervening long depression, or furrow, in the driest seasons."[9]

The efficiency of the heavy plow also encouraged communities to commit to clearing more of the forests, a topic I return to in the next chapter, as water-powered sawmills became more common.

The heavy plow by itself was not the only improvement. In combination with larger teams of oxen (and later horses, when they were drafted into service) and plows mounted on two wheels, farmers in the northern climes found they could sow much more seed and yield more varied crops if they pooled their resources. The heavy plow was not a tool that any one family of farmers or even those of a small village could afford. More often, the machine was a public investment of the early communities, and its adoption required the careful rethinking of property boundaries and how fields should be managed so that everyone could benefit. In the early Middle Ages, groups of free farmers were able to build mills collectively. With the rise of feudalism beginning in the eighth century, however, they lost this freedom. Large estates, owned either by the monasteries or lords granted lands by their kings, seized what they had not already acquired by purchase.

Over time medieval agriculture evolved into something decidedly different from that of ancient Roman practice, when farmers rotated only two crops. A farmer with two fields would let one lie fallow, unused, every other year in order to let the course of nature restore the soil after a crop yield. This included letting the livestock manure the grounds at their leisure.

By the 700s CE, farmers working for the growing monasteries began to adopt a three-field approach, in which the arable land was divided into three equal fields. During the course of one year, one field would be used for a winter crop like wheat or barley, the second field would be sown in the spring with oats or another spring crop, and the third field was left fallow until the next year. Then the farmer would use the fallow field for his winter crop while sowing spring seeds in the previous year's winter field and allowing the previous year's spring crop field lie fallow. In this way each field got a break—a restorative season as it were—every three years.

Among the advantages of this system were that a larger proportion of the farmland was available for cultivation. Having two different crops in two different seasons also offered some protection against the failure of any one crop. And if you used horses to haul the plow, as more medieval farmers did over the centuries, planting oats in the spring provided their feed. The yield from cereal cultivation depended on the effectiveness not only of letting a field lie fallow for a year, but also applying livestock manure and plowing it during this period.

The emergence of the horse as the primary draft animal is another key element of the transformation of agriculture in the later Middle Ages, one that neither the Greeks nor the Romans had ever seen as a need. Like the other cultures of the eastern Mediterranean, the ancients

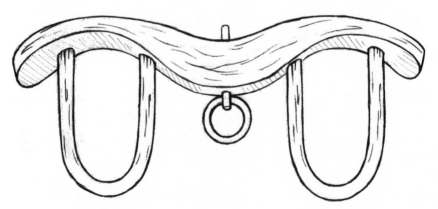

Medieval ox yoke for two oxen. ILLUSTRATION BY RYAN BIRMINGHAM

Horse collar. ILLUSTRATION BY RYAN BIRMINGHAM

relied on oxen for farming and mules and donkeys for pack animals. Horses were primarily used for the military—drawing chariots and later as cavalry after adopting the stirrup, which offered mounted soldiers a more formidable means of attack—but not as draft animals.

A key reason why horses were, for so long, not considered for plowing was the inefficiency of the yokes and harnesses employed for oxen. With their high shoulders, hump at the top of the back, and thick hide, the oxen could bear the weight of a wooden yoke around their neck and shoulders with a harness to help keep it in place; the pressure as they pulled distributed evenly, with most of the force forming around its shoulder blades, or withers as they were called.

On a horse, with its longer, higher neck and narrow withers, the yoke and harness were more likely to constrict the animal's windpipe and choke it. This was a problem, as J. G. Landels writes, that was not really solved until the 800s, when the yoke was dispensed with and the throat harness was replaced by a stiff collar, which put the pressure of pulling on the shoulders and chest of the horse.

In addition, not long after—toward the end of the 800s—horseshoes were adopted and quickly provided further advantages, as John Langdon points out. For one thing, horseshoes improved the draft horse's overall endurance, especially where the cold climate of northern Europe was a particular challenge. Unshod horses were apt to go lame, their hooves soft and easily worn down. In addition, iron horseshoes provided further traction on the road and in the field. Langdon points out that horseshoes were already known in the Roman era, both as a hippo-sandal, a kind of slipper made of metal, which was strapped to the horse's hoof with leather, and the more modern version of a bent strip of iron nailed to the horse's hoof.[10] But they seemed to vanish from the historical record after that until the ninth century. "It's difficult to know," he adds, "whether this is due to the inadequacies of the sources or whether an actual break in the use of horseshoes occurred. In any case, when they reappear in the documents, their adoption seems to have been rapid; by the eleventh century horseshoeing was an almost universal practice."

Once harnessed in this way, a horse-drawn plow was capable of turning over a field much more quickly than the ox-driven plow. This was a consideration that had to be weighed against its expense. And although horses did take on a greater percentage of plowing by the later Middle Ages, they never entirely replaced oxen because they required higher maintenance (perhaps like the more expensive farming tractors today). Horses need to feed on oats, which must be grown or bought, whereas oxen can survive on straw and grass. Like other ruminants such as cows, sheep, goats, and camels, oxen have digestive systems that are slower but far more efficient than those of equines. An ox can consume one and a half times as much common grass and straw as a horse and draw a higher proportion of the protein from what it eats.

The upshot of these three developments in tandem—the adoption of the heavy plow, the more efficiently harnessed and shod horse, and the institution of three-field crop rotation—was that by the year 1,000, the

European agricultural industry was well on its way to surpassing that of ancient Rome.

According to Jean Gimpel, in his book *The Medieval Machine*, between the eleventh and thirteenth centuries, the average crop yield increased "from approximately 2.5 to approximately 4. That is, for every measure of grain sown, the harvest yielded 4 measures. This meant that the portion of the harvest to be disposed of by the producer doubled—from 1.5 to 3."[11] In Gimpel's view, the plow was largely responsible for launching no less than an "agricultural revolution" in the later Middle Ages, a view also espoused by other mid-twentieth-century scholars like White. But more recent historians have cautioned that such a view overstates the case. "Most of the improvements in productivity which occurred in later medieval agriculture can be attributed to more intensive application of labor than to technological improvements," according to Adam R. Lucas.[12] Even contemporaries of White and Gimpel, like Duby, pointed out that without healthy draft animals, no plow was going to be as efficient as a dozen men with spades. In addition, many of the surviving records from estates in the eighth and ninth centuries—again, those operated by monastic orders—show that most of the farming tools in inventories were still made of wood and meant for the individual laborer rather than plow teams. So the question of a rapid "revolution" seems

Farmer with yoked oxen team. PHOTO COURTESY OF SUSAN SCHIBANOFF

indeed to be overstated, as it overlooks the long span of time during which the improvements in medieval agricultural tools took place.

And yet, to this day, in many parts of the world, developed and undeveloped, farmers plow their fields with the same teams of horses and oxen, the same heavy plow, and the same system of leaving one field fallow for every two sown. The fruits and vegetables that arrive in your local supermarket—especially those that are out of season during the northern winters—were probably grown by farmers using the same tools that were adopted and perfected by the farmers at the dawn of the second millennium. Although not perhaps revolutionary in the modern sense, as Lucas concludes, "technological developments in some regions of medieval Europe during certain key periods was significant, impressive, and perhaps in some instances might even warrant the appellation, 'revolutionary,' if suitably qualified."

## Chapter Two

# Harnessing Nature's Power

*Early Islamic societies appear to have developed a number of important new techniques for delivering water to run water-powered devices. They also appear to have provided a conduit for a number of Chinese (and possibly Roman) innovations in milling technology that were later adopted in medieval Europe.*

—Adam Lucas[1]

Today, it's ironic that "going green" is seen as a radical departure from the fossil-fuel-based economy we take for granted. But fossil fuels only have been mined and exploited on an industrial scale for the past two and a half centuries of civilization. For the vast majority of humanity's history, human beings exploited much simpler and more accessible natural sources of energy to support themselves—chiefly water and wind. That we are poised for a return to harnessing more energy from these sources today might have cheered medieval technicians if they'd been able to look into a crystal ball and see the future.

It's common to assume that technology evolves in a linear, progressive direction, starting with the simplest applications and proceeding to more complex models as humans improve their tools over time. For example, it was generally thought that the horizontal water mill for grinding grain into flour was developed first and that the more ambitious and industrial vertical water mill—with gearing to provide power directed at right angles to the wheel rotation—followed. Both of these waterwheels were inventions from antiquity that would be adopted for much wider application during the Middle Ages.

The horizontal water mill was certainly easier and cheaper to build. It was also better suited to hills and mountainous areas with limited water supply. In such settings, a modest cascade with a high head could be directed through a chute onto the oblique paddles of the wheel. Currently, there is no archaeological evidence of these models from classical antiquity, though there is written evidence. The earliest physical remains of their use in medieval Europe come from finds dating to the seventh century CE, although scholars believe they could date back to the fourth century CE.[2]

As for the more complex vertical waterwheel, scholars lately have come to agree that it seems to have been invented earlier, about the middle of the third century BCE, and to have proliferated more widely throughout the ancient world before the collapse of Rome than was previously thought. This was not an easy history to reconstruct. What written testimony exists from classical antiquity is sparse and not specific enough to establish a date of invention: we have an epigram written by Antipater of Thessalonika; a technical description by the Roman architect Marcus Vitruvius Pollio; and a brief aside by the geographer Strabo, all of these coming between 25 BCE and the beginning of the Common Era. They all treat the water-powered mill as something "new"—but without providing any more solid background information about its origins.

Antipater wrote an epigram in his *Palatine Anthology*, paying tribute to what was almost certainly an overshot vertical millwheel.

> *Cease from grinding, you women who toil at the mill; sleep late, even if the crowing cocks announce the dawn. For Demeter has ordered the Nymphs to perform the work of your hands, and they, leaping down on the top of the wheel, turn its axle which, with its revolving spokes, turns the heavy concave Nisyrian mill-stones. We taste again the joys of the primitive life, learning to feast on the products of Demeter without labor. (IX, 14)*

Besides giving a vivid poetic portrait of a mill wheel being turned by water nymphs, Antipater also betrays the fact that most hand milling was done by women at the time. More curiously, considering that he wrote from a pagan tradition, he nevertheless seems to harken back to Edenic times, before the biblical Fall, when humans did not have to work by the sweat of their brows.

Vitruvius provided more technical detail on the vertical water mill in the very brief chapter 5 in book X of his *De Architectura*:

> *Water mills are turned on the same principle, and are in all respects similar, except that at one end of the axis they are provided with a drum-wheel, toothed and framed fast to the said axis; this being placed vertically on the edge turns round with the wheel. Corresponding with the drum-wheel a larger horizontal toothed wheel is placed, working on an axis whose upper head is in the form of a dovetail, and is inserted into the mill-stone. Thus the teeth of the drum-wheel which is made fast to the axis acting on the teeth of the horizontal wheel, produce the revolution of the mill-stones, and in the engine a suspended hopper supplying them with grain, in the same revolution the flour is produced.*

Although the term "drum-wheel" was used for what would later be known as a cogwheel, a rotating gear cut with teeth, here Vitruvius correctly follows the path of power from the waterwheel's rotating vertical axis as it is transferred at a right angle to a rotating horizontal shaft reaching upward to power the grinding of the millstones in a separate chamber above the waterwheel.

Finally, in his *Geography*, book XII, chapter III, Strabo hails the water mill as a noteworthy part of the newly built city of Cabeira (near the modern-day city of Niksar in northcentral Turkey): "Now," he writes, "this city [Magnopolis] is situated in the middle of the plain, but Cabeira is situated close to the very foothills of the Paryadres Mountains about one hundred and fifty stadia farther south than Magnopolis, the same distance that Amaseia is farther west than Magnopolis. It was at Cabeira that the palace of Mithridates was built, and also the water-mills; and here were the zoological gardens, and, nearby, the hunting grounds, and the mines."

Apart from an allusion to the water mill by Lucretius, there are no other mentions of the technology in the surviving literature from antiquity. Nevertheless, this simple means of waterpower was in one sense a downward extension of a potter's wheel, so it's theoretically possible that the horizontal waterwheel has been known since the third or fourth millennium BCE.

As for physical remains, historian Örjan Wikander notes that to date there is no certain archaeological evidence for horizontal-wheeled mills in classical antiquity before the late third century CE. But from that era, the remains of two remarkable triple-helix turbines have been found, discovered in Tunisia, neither of which could be described as simple machines, though they were in principle horizontal-wheeled drives.[3] Until their discovery in the late 1970s, historians believed that turbines had not been invented until the sixteenth century.

The Tunisia waterwheel was set up at the bottom of a cylindrical well filled with water, which, according to Wikander, entered at a tangent through an inflow channel. In effect, it generated a rotating column of water, constituting a genuine turbine.[4] However ingenious this was for the time, no other such models were built and the turbine was not re-created until the 1500s in Renaissance Europe.

But the remarkable fact is that, in the space of a few centuries, we appear to have found three very different specimens of water-powered mills, and the simplest was by no means the first to be invented. One of the reasons for this assessment is that the components of both the horizontal and the vertical water mill already had long been known and used for a different application: lifting water, a task that was even more immediate to the needs of farmers than that of grinding cereal grains. The first bucket waterwheel, or noria, as it later became known in Arabic, was first used in ancient Egypt and dates to the third century BCE. They were known in China by the second century CE, if not earlier, and were widely used throughout the Middle East by the fifth century CE in order to irrigate crops.

Imagine a Ferris wheel mounted with water buckets rather than passenger seats and you can understand how the bucket wheel works: from a water source such as the Nile, the waterwheel, driven by the reliable force of the current, drew buckets up from the flow. When they reached the top of the wheel, a mounted spoke or lever tipped them over so that they emptied into a chute that would then flow down into troughs that could be directed into the fields for watering crops.

Not all rivers and streams have a reliably strong current, because the water flow, especially dependent on the winter and spring melting, changes with the seasons. A smaller noria for a single farmer could be turned by hand or by foot like a treadmill.

For use by a larger community, water was most efficiently raised by the power of a mule or other draft animal pulling a horizontal draw-bar in a circle centered around a vertical shaft, the base of which was fitted with a gearwheel that could in turn drive the horizontal shaft of the waterwheel. In this way, horizontal rotation of the drawbar from a vertical axle translated power to a horizontal rotating shaft to power the waterwheel.

The noria. ILLUSTRATION BY RYAN BIRMINGHAM

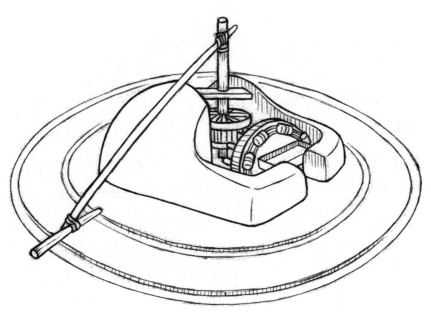

Gear-driven noria. ILLUSTRATION BY RYAN BIRMINGHAM

Here in this one device, the ancient noria, farmers possessed all the components needed to create a completely different machine: a water-powered mill that could grind grain. It's not a coincidence that the first water-powered mills initially appeared around the same time as the noria. But in the case of both the horizontal and vertical mills, the power worked in reverse—instead of animal-powered rotation to drive a water-lifting device, a water-powered wheel drove a shaft with enough power to rotate a millstone. (The Romans used various kinds of wheels for mine drainage, including the conventional noria.)

In the case of the horizontal water mill, no gearing was necessary. The upper stone of the quern, into which grain was poured and ground, could be attached directly to the shaft from the wheel below. Water was directed from a river or stream through a chute, or penstock, to drive the rotation of the wheel. For a mill of this type, the waterwheel was housed beneath the quern, with the shaft reaching upward (and through to a second floor), where the millstones were set up and the flour grains from the grinding were collected.

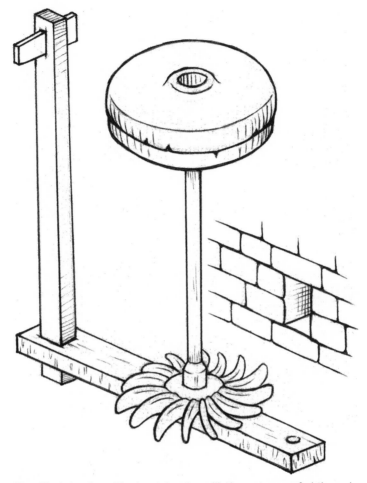

Simplified drawing of horizontal water mill: the water was fed through a chute in the wall, and the mill was mounted on the vertical shaft that powered the millstones on a separate floor. ILLUSTRATION BY RYAN BIRMINGHAM

In contrast, the vertical waterwheel's horizontal axle required another component, a shaft at a right angle to which it could transfer its energy to rotate the quern. This was achieved by cogwheels interlocked by the pins or teeth set around the circumference of a cogwheel at the ends of each shaft. The vertical water mill uses all the key components already employed by the noria: rotary millstones, or querns, for grinding grain (as we saw in the last chapter, they were already in use with the rotary hand mill); right-angle gearing for translating vertical rotational force to horizontal force; and finally, the waterwheel itself. These last two tools were also already being used for manual and animal-powered water lifting.

A major challenge for both kinds of mill—and in particular for the vertical water mill—was how to ensure the water flow would be available and reliably strong enough to drive the waterwheel. Given the vagaries

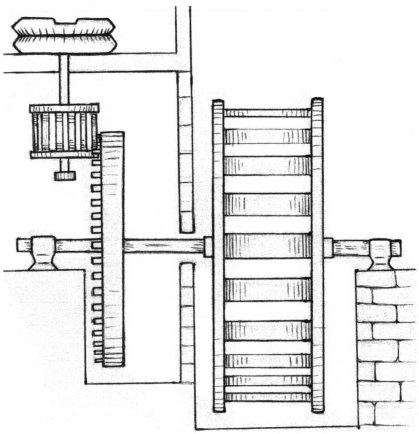

Components of the vertical-wheeled water mill: the mill wheel on the right drives the horizontal axle, which in turn rotates a toothed gearwheel that turns the vertical axle, driving the millstone above. ILLUSTRATION BY RYAN BIRMINGHAM

of the seasonal changes in rivers, it made sense to devise a means of regulating the water channeled toward the millwheel in a way that would free the mill from dependence on the river's seasonal rising or falling water level.

One way, which appears to have been common for small mills in medieval Britain and Ireland, was to build a weir, a dam of stones, across the river to raise the water to a reliable level. From this side of the pooled water, a wooden sluice channeled water down a chute or canal directly to the mill. The sluice could be equipped with a gate that in essence enabled the millers to turn the flow on or off. This also allowed maintenance and repairs on the mill wheel to be done much more conveniently than in waterwheels built directly into the natural flow of the river.

More ambitious communities would dig out and build an entire millpond that could be used to feed multiple mills at the same time. A third approach, most common in Mediterranean Europe, was called the tank mill, or *molino de cubo* in Spanish (known in Arabic as the *aruba*). Water was collected in a tank built of stones or bricks, and when sufficient head had built up, it released into a penstock feeding the water directly to the waterwheel.[5]

Once water could be harnessed in a more varied fashion, millwrights could also experiment with how the waterwheel could be driven most efficiently by it. There were three different solutions for the vertical waterwheel: the undershot wheel, the breast-shot wheel, and the overshot wheel.

The undershot wheel was submerged only partially to let the flow of water drive the wheel from the bottom. This model worked well when the mill could be built directly in a strong river's current, but it could also be fed from a chute. It relied on the force of the water only and worked against gravity, which diminished its efficiency in comparison to the overshot wheel. With the overshot, water was targeted to the top of the wheel from a chute or penstock so that the overshot was driven by a combination of the force of the flow as well as the force of gravity, thus it was more efficient than the undershot wheel. The breast-shot wheel, as its name implies, was a kind of intermediate design, wherein the water was directed to the middle of the wheel, but it was the least efficient and least utilized of the three models.

The advent of the vertical prior to the horizontal water mill might seem puzzling at first, given its use of gears. But since the first century BCE, the Greeks routinely had used gearing in water clocks and toys built for public display. More spectacularly, a complex arrangement of gearing was utilized in the Antikythera mechanism, which was built around this time. Retrieved from a shipwreck in the early 1900s, the mechanism has been revealed to be an astronomical computing device that was designed to show the position of the sun and the moon (including dates for eclipses) and that of the planets for years at a time.[6]

It may have been apparent from the beginning that right-angled drives in vertical-wheeled water mills were markedly more efficient than the single-shaft drives in horizontal-wheeled water mills. But at the same time, for smaller rural communities or even for a single farm or estate, the horizontal water mill was cheaper to build and maintain and so more likely to proliferate, as it seems to have done after the collapse of the Roman Empire. As population centers grew again, however, the more powerful mills returned. Castile in Spain, for example, saw a massive transition from horizontal to vertical water mills in the twelfth century.[7]

As to its ultimate origins, Greece, China, the Near East, and Palestine have all been suggested as places the horizontal mill most likely first appeared. Most recently, the remains of one horizontal water mill were discovered on the Crocodilion River in Palestine and have been tentatively dated between 345 and 380 CE.[8] The aforementioned turbines, based on the gearless horizontal water mills discovered in the Chemtou and Testour regions of Tunisia, also have been dated to the same period. Thus, for the time being, the physical evidence points to North Africa or the Middle East during the first half of the fourth century CE as the birthplace of the horizontal water mill, although this assessment could change with new discoveries.

What seems to be more certain is that the more complex vertical-wheeled water mill dates at least three centuries further back on the time line. As Lucas notes, "Both its relatively late appearance and the fact that the earliest types appear to have been the most complex places a serious question mark over the commonly made assumption that technological development usually proceeds from the simple to the complex."[9]

Again we see the suggestion that demand drives technology as much as ingenuity. As Lucas points out, the social and economic environment in which the milling technology emerged was one of significant urban expansion in societies that were just beginning to commercialize the art of baking bread. At the start of the Common Era, Rome's population was close to a million people. "Bread and circuses" was more than just a slogan; the mass production and distribution of bread to the poorest families in society was a key factor in maintaining order and social stability, and this in turn would have encouraged a proliferation of mills throughout the regions dominated by Rome. By the time the empire began to collapse, milling technology had become common even in the lands most distant from Rome (a horizontal water mill has been documented from fourth-century Ireland). And as we saw in chapter 1, there is more and more evidence to suggest that the technology never entirely disappeared, but rather became the seedbed for the eventual rebirth of industry in the Middle Ages.

At the outset, adoption of the empire's milling technology throughout Europe during the early Middle Ages was largely egalitarian in the different kingdoms and microstates on the continent—first come, first served, as it were. The common folk made do with the milling technology that best fit their locality economically. Only later as the feudal system was established and the church's influence spread with the vast estates of the monasteries did ordinary folk find themselves forced to give up their peasant mills to their new lords and facing the prospect of paying fees to have their grain milled on the lord's manor or on the estate of the monasteries. This was not a welcome development for the poorest communities, and I return to the topic in a later chapter.

But during the early Middle Ages, Italy quite naturally continued the milling traditions that had served the Roman economy. By the end of the 500s CE, the country was dominated in the north by the Lombards, while the central part of the country was ruled by the popes of the Catholic Church. The rest of the southern peninsula was under the control of the still-thriving Eastern Roman Empire based in Constantinople. In all three regions, mills were commonplace, and relatively weak oversight in comparison to the earlier Roman stewardship led to the wide-scale adoption of horizontal-wheeled water mills in the poorer and rural communities, while the more sophisticated vertical mills dominated in larger cities and on the major rivers.

To the north in Britain, the chaos that followed the departure of the Romans in the fifth century included the invasions of the Anglo-Saxons into the territories of the native Britons. But unity among the people who would become the English was only fleetingly achieved by the ninth century under King Alfred before the Vikings invaded. It would take another century for Alfred's successors to reconquer what had been lost to the new invaders, but only the briefest period of stability was enjoyed before the middle of the eleventh century, when the Norman invasion took place in 1066. Through all of that, the use of mills by the people working the land appears to have continued.

The famous *Domesday Book*, which William the Conqueror ordered to be drawn up so that he would have a detailed listing of all the lands and properties under his authority, was completed in 1086. According to its listings, there were more than six thousand water mills in the country held by the manors of his various lords and dukes. What's astonishing is that we have no record or physical evidence of water mills in Britain during the previous period, between the early fifth and late seventh centuries. And only a few sites have been confirmed during the period leading up to the Norman conquest, although there are documents in the form of charters referring to the existence of new mills in the eighth and ninth centuries.

It would appear that there were two different periods of mill expansion in England, with the development away from the vertical-wheeled mills from the Roman period toward the horizontal-wheeled mills between the sixth and seventh centuries, then a further more dramatic expansion and return to the vertical-wheeled mills starting from the time of King Alfred in the late ninth century.

Meanwhile, to the west of Britain, although the Romans never ruled Ireland directly, it seems that their technology arrived along with general trading that took place between the island and ports in the Mediterranean. Thanks to the survival of medieval Irish manuscripts, historians have determined that milling technology predated the arrival of the

Christian monastic orders, which helped spur the introduction of industrialized milling elsewhere on the continent.

One early Irish compilation of laws dating from the beginning of the seventh century, for example, provides evidence that water mills were already so common that landlords could seize them from people who owed money or rent. A later tract from about 700 CE regulated eligibility for mill ownership according to one's rank in society, so that, for example, a farmer of smaller property could not own as large a share in a mill as a farmer of greater stature. Another legal tract from around the same period, between the sixth and seventh centuries, contains the earliest terms for the components of horizontal mill wheels ever used in the vernacular of any European language. Still another legal tract is a first of its kind in medieval Europe for discussing the running of water mills as a specialized craft in its own right.

This textual evidence is also backed up by archaeological findings. Until the present, archaeologists in Ireland have uncovered forty water mill sites dating from between 340 and 1150 CE. A standout is the mill site discovered in Killoteran, County Wexford, which could be as old as the fourth century CE. If the dating assessment holds up, it would mean that milling was introduced to Ireland at the time that the Romans still occupied Britain. It appears to have been a horizontal-wheeled tide mill. Based on the sheer number of mills between the seventh and eighth centuries, it also appears that Ireland's milling industry easily rivaled that of the European kingdoms, at least partly due to the fact that it was relatively free of invasions in contrast to Britain and the continent.

From the northwest of Europe, we turn now to the Near and Far East, as milling technology made its appearance in China at about the same time or shortly after it emerged during the Roman Empire. It's also one of the rare instances in which a technology emerged in the West and was adopted from it in the East. As we see in later chapters, many of the inventions adopted and expanded in the Middle Ages had their origins first in China and the Islamic Middle East before being transmitted to the West.

It was during the Han Dynasty, between 202 BCE and 220 CE, that water-powered technology made its first appearance in China. Gearless vertical waterwheels with camshafts were employed to power trip-hammers for the hulling of rice and for powering forge bellows, both around the start of the Common Era.

It is not until the early 600s CE during the Tang Dynasty of emperors that definite textual evidence of geared vertical-wheeled water mills like those being used in the West is available. As in the West, too, it seems that the more complex vertical-wheeled mills were owned and managed by higher-status social groups, such as wealthy merchants, Buddhist monasteries, and high-ranking officials in the imperial bureaucracy. So important, in fact, did water mills become to all levels of the Chinese economy that the Tang emperors appointed two commissioners to oversee the regulation and rights of ownership of milling. Strictly speaking, only four powerful social groups were granted the right to own and operate mills in China: imperial concubines, high-ranking officials in the imperial bureaucracy, Buddhist abbeys, and wealthy merchants.

The Silk Road, one of the trade routes established between China and the Mediterranean by the Han emperors in the last century BCE, was also instrumental in bringing much of the technology that the newly emerging emirates of Islam would adopt in their conquest of lands from Persia in the east all the way to Spain in the west. The sites of twelve horizontal-wheeled water mills have been found in Oman on the Arabian Peninsula dating between the eighth and tenth centuries. In Iraq and Iran the remains of thirty-one mills have been dated to between the seventh and thirteenth centuries.

Islamic engineers seem to have been particularly ingenious in adapting the vertical-wheeled water mill to stronger sources of waterpower. Ship mills, for example, were built on great barges that could be anchored in different currents. Mills were also constructed as permanent parts of bridges spanning powerful rivers, the earliest record of one being from Cordoba in Islamic Spain during the twelfth century CE.

While European millers were designing wooden sluice gates and chutes to channel water from mill ponds, Islamic engineers in Persia were building great stone dams through which pipes could feed water directly to horizontal-wheeled water mills. As Lucas writes, they went further, building water mills "within underground irritation tunnels, or *qanats*, in order to exploit the flow of water in the tunnels." A Persian invention, qanats were ubiquitous throughout the East and the West. In Muslim Spain alone, qanat installations ranged from huge, monumental tunnels with fifty airholes to "backyard" versions with only two or three. Ownership and maintenance of the mills in Islamic lands was more varied than in feudal Europe and imperial China, and vertical-wheeled mills throughout the Islamic world were more rapidly employed for other purposes besides grinding grain: sugarcane maceration, rice husking, powering forge bellows, ore crushing, fulling cloth, sawing timber, and prepping pulp for papermaking. It would not be long before European mills also expanded their scope of operations.

What does seem to have been unique to Europe—and indeed may have originated in Ireland—is the development of coastal tide mills. A number of vertical- and horizontal-wheeled water mills were built on the coasts to take advantage of the swelling of inlets and rivers from the ocean, which could be channeled directly to mill wheels and provide power for between six and twelve hours of work.

The mills in use on the coast were no different than those built to exploit rivers inland. But the means of capturing and storing the water required alternatives to the weirs, dams, and millponds that millers employed for river-based water mills. As Lucas describes it, a tidal-based millpond, presumably excavated in the marshes, would be designed to allow the incoming tidewater to enter through a so-called sea hatch, a swinging gate that automatically closed once the level of water in the sea-water pond had risen high enough to force it to close by natural pressure. A sluice gate close to the mill then could be opened at will to channel the water to the millrace, where it would be directed onto the mill wheel.

The previously mentioned Irish mill dated to the fourth century CE that was discovered at Killoteran may have been a tide mill. Dated more specifically to 630 CE were two tide mills, one a horizontal-wheeled mill and the other a vertical, both on Little Island on Lough Mahon in Cork.[10]

In the south of England, off the River Test in Southampton, one can still visit the site of the only surviving tide mill in the United Kingdom, the Eling tide mill, which was first referenced in the *Domesday Book* in the eleventh century and continued in operation—with renovations and rebuilding—until the early nineteenth century. It was restored in the 1980s and is now a tourist museum, but the mill is operational and can mill grain for up to twelve hours per day.

However the technology spread, there is textual reference to a tidal mill from Basra in Iraq built on the Shatt al-Arab river near the Persian Gulf and enthusiastically described by the geographer Al-Muqaddisi in a book dated to around 900 CE. Tide mills were also built along the coasts of Portugal, Spain, the Netherlands, and Flanders, among other locations in Europe over the next two centuries.

The great drawback of tide mills was their vulnerability to flooding during coastal storms. Their use seems to have peaked between the eleventh and early thirteenth centuries. With the population growth and continued demand for milled grain, other approaches to milling had to be developed, especially in the low-elevation regions of northern Europe, where rivers and brooks were either too few or simply not reliable enough to generate sufficient power for water mills. And so millers began to exploit another power source that had not yet quite caught on in the southern regions of Europe or the Mediterranean: windmills.

Here, once again, there seems to be no direct connection or transfer of technology between the first windmills developed in the Middle East and those Europeans later adapted to their windy northern climes.

The horizontal windmills of Persia date from between the fifth and the ninth centuries CE. They were built to operate inside tall, partially gutted mounds of earth and stone in the hillsides and always faced the same direction—north—whence the winds were most prevalent in that region. From a distance they must have looked like giant stone cylinders with one side cut away to fit the mast and sails, which spun around inside from the force of the winds. Some scholars believe these horizontal windmills were first inspired by Buddhist flywheels in India, but there is no concrete evidence. The sails were mounted on a single mast or post that rotated through the force of the wind and ran directly down into the housing of the mill where it drove the millstones.

In contrast, the windmills that developed in Europe appeared in the eleventh century and were based on a clever inversion of the already existing vertical-wheeled water mill. There were two types: the post-mill

Persian windmill. ILLUSTRATION BY RYAN BIRMINGHAM

and the tower mill. Both improved upon the horizontal windmills of the earlier era by allowing the mill to be rotated in order to exploit the winds whatever their direction.

The post-mill was easier to build but limited compared to the tower mill. At its most basic, the post-mill was like a giant treehouse; the entire structure—four walls and a roof—was built to revolve around a central post planted deep into the ground like a tree trunk. The vertical wheel with sail stocks was mounted on the exterior upper floor of the house on a horizontal shaft with a brake wheel inside. A gearwheel at the inner end of this connected to and rotated a central vertical shaft running down to the millstones on the lower floor. Access to the millhouse was via ladder. The entire windmill could be rotated from the outside by means of a sturdy tail pole attached to the base, either by hand or by a draft animal. Subsequent designs reinforced the central post with quarter struts some-

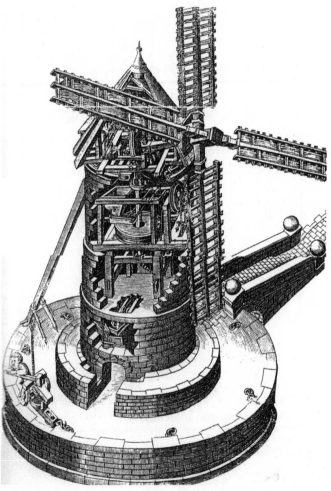

Tower mill with tail pole and rotating turret. ILLUSTRATION BY AGOS-TINO RAMELLI, FROM HIS BOOK *THE VARIOUS AND INGENIOUS MACHINES OF AGOSTINO RAMELLI*, 1588. PLATE 132 IN THE DOVER EDITION, 1976, TRANSLATED BY MARTHA TEACH GNUDI AND EUGENE FERGUSON

times rooted in a foundation of crossbeams or a stone wall to make it even sturdier.

The drawback of the post-mill was its vulnerability to changing wind directions. The brake wheel was not sufficient by itself to stop the rotation of the sail stocks. The mill had to be turned ninety degrees out of the wind first, and any error in turning too far could expose the sail stocks to the winds from behind, which could destroy the mill altogether.

The tower mill, as its name suggests, took the post-mill design a step further by mounting the entire millhouse and sail stocks as a rotating turret built on top of a stone tower. In addition to fortifying the windmill more strongly against the elements, this approach also encouraged expanding the construction of windmills onto preexisting stone structures, such as castles and bridges.

Early models of the tower mill required a tail pole, like the post-mill, to rotate the windmill from outside the tower. Later designs mounted the tail pole on wheels and a track that could be more easily drawn by a horse or mule. By the seventeenth century, the turrets were mounted on internal tracks and could be rotated entirely from within.

The tower mill was considerably more expensive to build, and during the era of the most rapid growth of medieval windmill technology—the eleventh to fourteenth centuries—the post-mill remained the most common. By the end of the 1200s, the post-mill could be found everywhere in France and England where waterpower was limited and steady winds were reliable. They were also adopted in the Netherlands, Denmark, Bohemia, Poland, and Sweden before the first half of the fourteenth century.

Because of their efficiency and reliability, the windmill eventually replaced the tide mill as population growth increased the demand for grains right up until the onset of the Black Death in the mid-fourteenth century. And although windmills never replaced water mills, their design and rapid adoption throughout northern Europe constitute in the eyes of today's historians a genuine technological revolution.

# The Crank and the Camshaft

*Taken together, the results achieved during the last two decades are enough to show beyond doubt that water-power was applied in the Roman Empire to various purposes besides grain-milling. From a technical point of view, we are dealing with two different kinds of machinery: a simpler one operated directly by the continuous rotary motion of the water-wheel, and a more complex one requiring conversion to linear motion, achieved by the application of a cam and lever.*
—ÖRJAN WIKANDER[1]

OF ALL THE INVENTIONS THAT ARE COVERED IN THIS BOOK, THE HAND crank seemed to be the most elusive to historians until recently. And yet, in combination with the machines we have already seen, it would have a huge impact—especially on the expansion of mill-powered technologies in late antiquity and then in the late Middle Ages. The crank arrived in concert with many of the other components we have seen emerge from the last centuries before the Common Era, such as the waterwheel, the millstone, and the gearwheel, however faint its historical imprint appeared over the centuries.

At its most basic, the hand crank served as a handle attached at a right angle to a device designed for rotary motion, such as a shaft or a disk, to more easily operate it by hand. It appeared on the ancient lever mill, for rocking the top millstone back and forth. Then it was found on the rotary hand mill, which made life a little easier for the women and slaves charged with turning the heavy disk-shaped millstones, from which comes the expression, "the daily grind."

In China during the Han dynasty, encompassing the period from two centuries before the Common Era into the Common Era, the crank was employed to turn a rotary winnowing fan to help separate grain from chaff and to power a winch for drawing water buckets from a well. Both examples show how the crank could be used to convert rotary motion into reciprocal motion. In the case of the well, the crank is attached to a winch with rope for drawing a water bucket from (or lowering it to) the water below; the rotary motion of the hand-cranked winch converts to the rectilinear motion of the rope, drawing up or letting down the bucket. This convenient solution to raising water would soon evolve into more ambitious tasks.

In the West, the crank seemed like a logical addition to the invention of the gearwheel, credited to Archimedes in the late third century BCE. Although not certain, it seems likely that a crank was the chief means of manually operating the Antikythera mechanism, which ran on a series of internal interlocking gearwheels to simulate the motion of the planets on multiple external dials.

Beyond that period, for a time, historians had only a few tantalizing hints regarding how common the crank may have been before the Middle Ages. What appears to be a crank handle dating from the second century was discovered in Augusta Raurica, the Roman archaeological site in Switzerland. It consists of an iron rod about eighty-three centimeters in length with a bronze crank handle attached at one end. But it seems to be missing a counter handle on the other end. What device it was intended to operate remains a mystery, and it did not impress scholars once inclined to be skeptical of Roman innovation.

The lack of archaeological (and written) evidence for the crank is one of the reasons that many historians—especially until the end of the last century—embraced a kind of "stagnation" thesis about technology in the classical world of Greece and Rome.

This was partly inspired by the derogatory comments found in the writings of Plato, Aristotle, and other ancient writers, apparently looking down their noses about the merits of practical technology as not being as worthy of study or interest compared to the more "divine" sciences of philosophy and theology. As brilliant a mathematician and engineer as Archimedes was, he did not leave any written works about his inventions because he reputedly found the topic sordid. This suggests that, at a purely intellectual level, the ruling classes of the Greeks and the Romans thought craftsmanship and technology were beneath them. Coupled with a dependence on slave labor, it seemed as though any investment in the creation of labor-saving devices was discouraged as being unprofitable. According to this thesis, it was not until the later Middle Ages, when slavery was disappearing in the West, that a different attitude toward the natural order arose, promoted in large part by the ascendant Christian Church. This in turn finally inspired a new age of invention, a view adopted by many historians of medieval technology such as Lynn White Jr., Marc Bloch, and Jean Gimpel. It's a view that obviously appealed to Christian apologists anxious to counter the notion that Christianity was in any way anti-intellectual or discouraging of science and technology.

Over the last couple of decades, a new generation of historians has begun calling this view into question. As archaeologist Andrew Wilson notes, most of the general histories on the topic, though still in print, are several decades old and still subscribe in some form "to the broad outlook that the Classical world can take credit for relatively few new inventions; and that a number of social factors combined to retard innovation and inhibit the uptake and spread of new inventions."[2] This assessment began to change in the 1980s and 1990s, when more archaeological discoveries began to emerge that undercut the stagnation thesis.

One of the most remarkable discoveries, from 2007, revealed compelling evidence that by the early third century CE cranks were indeed adapted by the Romans to work in vertical water mills in order to power stone saws in Hierapolis, an ancient city in Phrygia, now the city of Pamukkale in southwestern Turkey.

On the sarcophagus belonging to an engineer named Marcus Aurelius Amminaos, archaeologists found a fading inscription and a sculpted relief of a sawmill. Written in Greek, the stoic words were translated as "M. Aur. Ammianos, citizen of Hierapolis, skillful as Daedalus in

Sarcophagus at Hierapolis with a relief of the sawmill. IMAGE COURTESY OF PAUL KESSENER, TULLIA RITTI, AND KLAUS GREWE

wheel-working, made the [the represented mechanism] with Daedelian craft [or 'with the skill of Daedalus']; and now I'll stay here."[3]

The relief is remarkable in that it clearly shows a breast-shot vertical water mill wheel (on the right side of the relief) being fed by a chute as well as a sequence of two gearwheels connected by rods to two different stone saws embedded in frames. Both saws are depicted as sawing halfway through slabs of stone. From the position of the gearwheels located side by side from a shaft connected to the waterwheel, the archaeologists inferred that the saws were powered by connecting rods that drove the saws back and forth from the rotary motion of cranks on two sides of a gearwheel powered by the water mill. A reproduction of how it most likely worked appears on the next page. Note that the crank fixed with the connecting rod is attached at the edge of the gearwheel's radius, offset from the center of the gearwheel, which allows it to push and pull the saw back and forth as the gearwheel makes an entire revolution.

A decade before this discovery, another Roman mill dated to the late second century CE was found in the town of Aschheim in Bavaria. Based on the reconstruction by the archaeologists, it was operated by a hand crank turned vertically. "A large cog turned by the crank was geared to a small pinion on the axle of the millstone, which therefore rotated much faster than the crank. This pinion . . . was not toothed along the rim but, like the standard large cog wheel employed by the millwrights of the ancien régime, consisted of short, thick pegs set vertically into the rim on

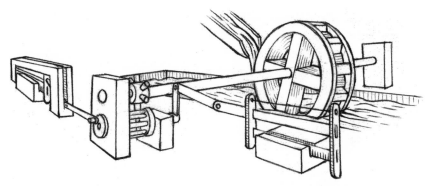

Drawing of the sawmill based on the relief carved on the sarcophagus. ILLUSTRATION BY RYAN BIRMINGHAM

one side of the wheel but without sticking beyond the circumference."[4] Thus, the crank seems to have made an earlier appearance than historians initially thought.

The cutting of stone into blocks or veneer slabs was already an old tradition by the time the crank was applied. According to Pliny's *Natural History*, the people of Caria, a kingdom on the coast of what is now southwestern Turkey, might have invented the art of stone cutting. As further proof, Pliny pointed out that the oldest marble slabs he knew of at the time were found at Halicarnassus, the largest city in Caria, in the palace of King Mausolus who died in 352 BCE.

It was not until the first century BCE that the style of covering walls with marble slabs was brought to Rome by Julius Caesar's prefect of engineers, Mamurra. He employed a manually operated saw, which was made of a wooden beam with supports on either end that were perpendicular to the axis of the beam. The saw blade was mounted on the inner side of the beam and kept in tension by means of a rope tightened on the back end of the beam. "The length of the saw and the distance between the saw blade and beam determined the size of the slabs and the blocks that could be sawn," according to the archaeologists at Hierapolis. For hard and durable material, the blade was toothless, the cutting of the stone achieved by sand. The saw acted by pressing on the sand within a very fine cleft in the stone as it worked back and forth.[5]

39

The crank was not seen again in Europe until the late Middle Ages, the late 1400s, when it was adopted for powering sawmills and piston pumps and for mills dedicated to grinding pigments for pottery glazing.[6] Although the crank and connecting rod ultimately replaced the camshaft in many mill operations, the camshaft remained the primary engine of the medieval milling industry through the fourteenth century.

Indeed, according to M. J. T. Lewis, "The marriage of the cam and the waterwheel, whenever and wherever it may have taken place, was as momentous a step as the marriage of the right-angle gear and the water-wheel, for it opened the way for a great variety of industrial mills."[7]

The cam—the key component of the camshaft—was known to ancient engineers. And, as was the case with the initial invention of gearwheels, cams were employed by their Greek inventors primarily as a means to power toys and gadgets that moved for show. The earliest example—known only through a reference by a Greek Egyptian author named Athenaeus of Naucratis—described a giant mechanical likeness of the nymph goddess, Nysa, created as part of a moving display in a procession celebrated by King Ptolemy II Philadelphus around the year 275 BCE in Alexandria. Twelve feet tall when standing, the figure was designed to rise from her seat, holding a bowl of milk, which she poured out as a libation (presumably into the empty cups of those accompanying her vehicle in the procession) before sitting down again, and, after a pause, rising to repeat the action. Historians have deduced that the wheels of the slow-moving vehicle may have powered the automaton with some kind of a cam on the axle connected to a lever or series of connected levers that controlled the figure internally.

The Chinese were the first to see in the cam a means of mechanizing a tedious but necessary task—the hulling of rice. As early as 290 CE, the Chinese employed vertical water mills to power trip-hammers for pounding the stalks bearing the raw grains so that they could be more efficiently winnowed for food preparation.[8]

We saw in the last chapter how the vertical-wheeled water mill drove a horizontal turning shaft, which, by means of a right-angled gearwheel, was able to rapidly rotate a vertical axle to power millstones for grinding grain. For hulling rice, the Chinese millwrights made a key adaptation to the horizontal shaft of the vertical mill wheel: they embedded care-fully placed fin-shaped protrusions, or cams, which could lift—and then drop—a series of trip-hammers that would pound stalks of rice. It was another ingenious way to convert rotary motion into reciprocal motion. In this case, it was the perpendicular protrusion of the cam from its rotating shaft that allowed it to push a corresponding hinged shaft with a hammerhead up a discrete distance into the air, after which it rotated away, allowing the trip-hammer to drop via gravity and pound its target. In the case of the Chinese mill, this consisted of stalks of rice laid out in a trough.

The same technology would eventually find its way to medieval Europe, where it would be adapted for pounding a number of different kinds of raw material, for example, in the process of fulling cloth. By

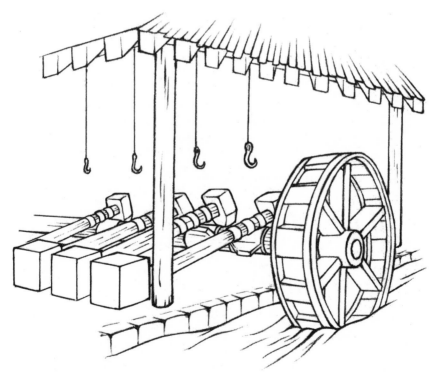

Chinese rice mill powering trip-hammers. ILLUSTRATION BY RYAN BIRMINGHAM

setting the cams along different lengths of the shaft and on different parts of the shaft's circumference, the engineers could alternate or synchronize the sequence of pounding.

As was the case with its counterpart, the crank, historians initially believed that the camshaft did not have an immediate impact beyond the culture in which it first appeared. A gap of centuries seemed to separate the first use of camshafts in ancient China with their appearance in Europe toward the end of the ninth century.

Recent evidence has been uncovered revealing that Roman mills employed trip-hammers for the purpose of metalworking in a fashion similar to the Chinese rice mills. As Örjan Wikander writes, at several mines dating from the first and second centuries CE in Spain and Portugal, archaeologists discovered a particular kind of stone anvil scarred with deep depressions made by ore-crushing stamps. The marks were

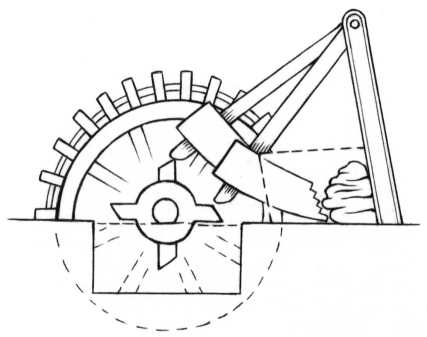

European fulling mill with trip-hammers powered by camshaft. ILLUSTRATION BY RYAN BIRMINGHAM

so regular that archaeologists concluded the pestles that formed them could not have been wielded manually. It is more likely that they were systematically raised and dropped within a guiding framework run mechanically—and almost certainly—by camshafts.[9]

The technology made its way to Europe most likely through the influence of Islamic engineering, which by the eighth century employed the camshaft in mills, both in Spain and elsewhere in the Middle East kingdoms of the Islamic caliphates. Although it's generally believed that engineers in the eastern kingdoms of the Middle East adopted Chinese milling technology that later found its way into Europe, it is perhaps possible that the Roman technology persisted in Spain after the collapse of the empire, where it was adopted by the Muslim invaders in the eighth century. The new rulers improved on the design as they employed water mills for purposes beyond forging iron. From there, the technology was likely transmitted into the rest of Europe as well.

Whatever the case, between the tenth and fourteenth centuries, non–grain grinding water-powered mills employing the camshaft would sprout all over Europe, with the largest concentration in the kingdoms that would become France. These new mills were built for a wide number of processes: pounding hemp for rope making, pounding tree bark for tanning leather, forging iron for building construction and for weaponry, crushing ores, fulling cloth, and sawing timber.

Between the years 770 and 1600 CE, the number of documented industrial mills in Europe currently known to historians totaled just under eleven hundred, with the largest share by far those of fulling mills in France and forge mills (between both England and France). The first fulling mills appeared by 1040 CE. Leather tanning mills appeared in 1138, and hemp mills by 1200. Timber mills were not recorded until 1300.

These last centers of medieval industry are noteworthy because historians have not been able to determine exactly how the earliest medieval sawmills were designed. Apart from a tantalizing diagram of a camshaft-powered sawmill in the notebooks of the architect Villard de Honnecourt (ca. 1235), there is no other evidence of how saws were driven in medieval timber mills until 1500, when cranks and connecting rods were employed instead of cams to power the reciprocal back-and-forth motion necessary for sawing.

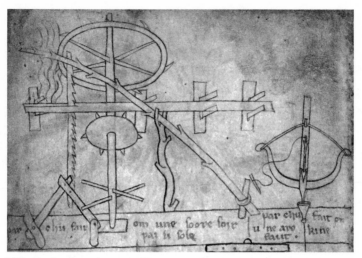

Villard's drawing of a sawmill. COURTESY OF THE BIBLIOTEQUE NATIONAL
DE FRANCE

Villard's design was probably sketched hastily, and it had the appearance
of a plan plausible for something like a toy rather than a water-powered saw.

But the general principle behind it addressed the challenge of convert-
ing the water mill's rotary motion into reciprocating motion—specifically
to drive the saw back and forth through logs of timber. In addition, the
machine also required a means of feeding the logs continuously into the
path of the saw. Villard's water mill featured a vertical saw blade drawn
downward by an elongated cam on the mill axles. The cam rotated downward to push on an angled guide bar attached at the base of the saw, which
itself is grounded by another horizontal guide bar, the two forming a kind
of jointed triangle with the one as the base and the other—receiving the
force from the cams—as the hypotenuse. The saw was attached at the top
to a strong sapling branch that acted as a kind of tension spring to draw
it back upward after the cam from the mill axle completed its rotation
and released the base of the saw guide bar. The sawing motion would be
repeated once the cams on the axles turned around and pushed down on
the guide bar again. A separate toothed wheel on the camshaft would feed
the timber continuously into the saw until it was completely cut.

Villard left no written notes with his diagram and the flimsiness of the sapling spring pole renders the whole contraption rather dubious.[10] This raises the question of whether the crank made an earlier appearance for use in medieval sawmills. In such a mill, perhaps one model could conceivably have entailed a horizontal camshaft driven by the standard vertical-wheeled water mill, but in this case one in which the cams triggered a geared crank to power a connecting rod, which could have driven a more formidable toothed saw.

When sawmills made their mark in the later Middle Ages, we know they accelerated what was already a huge program of deforestation taking place throughout central Europe that had begun in the eighth century when the population growth spurred the clearing of more and more fields for the purpose of farmland. This was not accomplished with only axes and plows. Controlled burning was also employed to quickly take down large sections of forest.

The issue of deforestation was not a new one in the Middle Ages. Humans began to change the landscape once the agricultural revolution took place ten thousand years before, and the struggle to make arable land out of the natural landscape was one of the biggest challenges facing every ancient civilization.

In his *De Rerum Natura*, written in the first century BCE, Lucretius lamented the faultiness of creation, the apparent poor design, and alluded to the thankless task of trying to refashion the earth so that crops could be planted.

*First, mark all regions which are overarched*
*By the prodigious reaches of the sky:*
*One yawning part thereof the mountain-chains*
*And forests of the beasts do have and hold;*
*And cliffs, and desert fens, and wastes of sea*
*(Which sunder afar the beaches of the lands)*
*Possess it merely; and, again, thereof*
*Well-nigh two-thirds intolerable heat*
*And a perpetual fall of frost doth rob*
*From mortal kind. And what is left to till,*

*Even that the force of Nature would o'errun*
*With brambles, did not human force oppose,*
*Long wont for livelihood to groan and sweat*
*Over the two-pronged mattock and to cleave*
*The soil in twain by pressing on the plow.*[11]

As the population grew between the seventh and thirteenth centuries—right up to the devastation of the Black Death—the forests of Europe began to shrink due to the demand for more arable land. And as we saw in chapter 1, the success of the heavy plow encouraged farmers to clear the lower lands with heavier, richer soils. Controlled burning not only cleared away the trees, but the ashes further nourished the ground prior to plowing and planting.

With the advent of the mill-powered saw during the later Middle Ages, the twelfth and thirteenth centuries, deforestation accelerated to the point of alarm. As Norman Cantor wrote in his book *The Civilization of the Middle Ages*,

> *Europeans had lived in the midst of vast forests throughout the earlier medieval centuries. After 1250 they became so skilled at deforestation that by 1500 AD they were running short of wood for heating and cooking. They were faced with a nutritional decline because of the elimination of the generous supply of wild game that had inhabited the now-disappearing forests, which throughout medieval times had provided the staple of their carnivorous high-protein diet. By 1500 Europe was on the edge of a fuel and nutritional disaster, from which it was saved in the sixteenth century only by the burning of soft coal (which, in turn, started air pollution) and the cultivation of potatoes and maize (Indian corn as fodder for cattle) that were imported from America.*[12]

Historian Teresa Kwiatkowska writes that scholars estimate the forests of France were reduced from thirty million hectares to thirteen million hectares between 800 CE and 1300 CE. Although up to 70 percent of the land in Germany and central Europe was believed to have been covered in forest in 900 CE, by 1900 only about 25 percent remained.

"The essential reason was an extensive growth of European population between 650 and 1350, resulting in a vast spreading of settlements in the forests of central and Eastern Europe."[13]

How was the camshaft utilized in mills dedicated to other processes that were less damaging to the environment? In the case of blacksmithing, there was a need to automate the driving of the forge bellows, which was designed to be worked by hand, and for that reason would have been limited in its size and capacity. A camshaft with longer, wider "fins" could, through lifting action, operate several larger sets of bellows at once, driving the air into the blacksmith's furnace as the cams rose and forced the handles of the bellows to close before releasing and then allowing the bellows to open and reinflate in time for the next revolution of the shaft.

For fulling mills, the mechanization was straightforward: what had been done by human feet, stomping on cloth in vats of water mixed with cow urine and manure, was taken over by recumbent trip-hammers. The process was likely first developed by Islamic fullers in Andalusian Spain or the Middle East as early as the eighth century, inspired either by late Roman technology or by similar mill technology transmitted via trade with China.

Fulling was the last step in a process of producing cloth from the wool of sheep. After shearing the animals in the spring, the raw wool was scoured and carefully combed to remove unusable clumps and tangles and then spun into fine yarn by hand or on a loom that worked by hand or foot treadle. These in turn were woven into long yards of wool. To be completely cleaned at this stage, the long yards of woven wool, up to thirty-two in an average "bundle," had to be pounded in the natural detergents just described: water with stale human and cow urine and some manure, both of which were rich in ammonia. "Fuller's dirt" was also an ingredient: a fine clay that contained silica and was ideal for degreasing and bleaching the wool. Fullers also used soapwort, whose natural chemicals, saponins, were also a key ingredient for cleaning. These were all mixed with water in the troughs for pounding. The process not only cleaned the wool of grime and impurities, but it shrank it by 10 to 20 percent in size, making the cloth tougher and longer lasting; it also gave the cloth a softer feel by the time it was woven into garments.

Recumbent trip-hammers in the fulling mills were powered by a horizontal shaft with cams lined up in staggered points along the shaft to lift the hammers into the air so that the natural force of their falling back into the trough pounded the cloth as it soaked in the detergent. The troughs were curved at the back so that the cloth would naturally turn over in the process of pounding, making it easier for the fuller to ensure that the entire cloth was completely treated.

Apart from the smell, the fulling mill must have been a loud place to work—the pounding of the hammers could be heard for miles around, and one wonders about the long-term effects on the fullers' hearing. It was also dangerous work, as the fuller needed to supervise the pounding hammers, and one slip near the splashing troughs must have led to serious injuries and even fatalities in some cases.

The smell from the urine-based detergents, though strong, was likely much less bothersome, as people of the time were already used to far stronger stenches from their farm animals, their open cesspits, and their unwashed selves, as bathing was far less frequent among the average European. "If the late twentieth century is scented with gasoline vapor and exhaust fumes," write the English historians Robert Lacey and Danny Danziger, "the year 1000 was perfumed with shit. Cow dung, horse manure, pig and sheep droppings, chicken shit—each variety of excrement had its own characteristic bouquet, from the sweet smell of the vegetable eater to the acrid edge of gut-processed meat, requiring the human nose of the year 1000 to function as a considerably less prissy organ than ours today."[14]

In total, more than 630 fulling mills are documented from the Middle Ages, making them by far the most numerous type of mill officially noted. This includes close to 250 from England, 200 from Wales, and almost 90 from France. There were 74 noted in Italy, and a comparative handful from Germany, Poland, Denmark, and Switzerland.

Most of the French mills date from the later Middle Ages. However, if one examines the dates when the mills were recorded as being in service, the mills dedicated to producing the more complicated products appeared latest in the Middle Ages. Whereas grain grinding for flour and for malts appears as early as 770, and the fulling mills appear circa 1040, mills for forging iron do not emerge until 1200 at the earliest. Paper mills appear around 1276. Mills for crushing ore are recorded in the early 1300s, and those for producing pigments for coloring clothing not until the late 1300s. Sawmills appear between 1300 and 1347.[15]

One of the more interesting developments from historical surveys comes from England, whose medieval milling culture has garnered more attention from historians than the rest of Europe to date. During the time of the Black Death (1348–1350), when the population was decimated by the plague, a large number of grain mills were converted to fulling mills. This was an easy and inexpensive adaptation, as the vertical-wheel water mills had all the gearing in place except for the trip-hammers.

The camshaft was the key to powering the mass production of cloth, iron, hemp, leather, and paper, all of which benefited from the repeated force of heavy pounding. Besides being a source of mechanical power, the cam also represented another kind of power—in the mind of humanity. As Jeremy Naydler argues, the camshaft represented the first example of machine programming in human history. By adjusting the cams on a shaft, the millwright could "program" both in what order and at what speed the mill's trip-hammers would operate. Naydler believes this blossoming of application to be an outgrowth of the Greek philosophy that medieval Europe inherited from the Greeks.

> First of all, we see how, within just a few centuries of the formalization of the rules of logic, the application of logical thinking to the construction of complex machines that substitute purely mechanical actions for human actions becomes a reality. Secondly, we may also notice that the functioning of the industrial watermill depended upon a primitive form of programming, and should therefore be regarded as a distant precursor of the computer.[16]

The water mill's role in producing paper can perhaps be viewed as another kind of precursor of the computer. Paper was certainly the result of a fortuitous conjunction of cam and gearing. And to that topic we turn in the next chapter.

## CHAPTER FOUR

# The Paper Explosion

*What Bacon also failed to note—perhaps because it was as obvious in
the seventeenth century as it is today—is that without paper, there
would have been no printing, one of many instances in which scholars
have lumped the pair together as allied technical advancements, with
paper usually getting the shorter shrift of the two, especially in the
impact they have had on the diffusion of culture.*
                                      —NICHOLAS A. BASBANES[1]

THE MASS PRODUCTION OF INEXPENSIVE PAPER HAD A PROFOUND
effect on late medieval society (between the mid-1400s to the early
1500s), more than any of the water mill–produced tools examined so
far. With the rise of the merchant classes, the demand for paper—for
accounting and bookkeeping, even prior to reading for education and
pleasure—would help to inspire the Renaissance. In a sense, this point
may serve as a contradiction to the technological fallacy discussed previously, for it seems beyond dispute that the widespread availability and
affordability of books—brought about by the success of the Gutenberg
printing press combined with the availability of inexpensive paper—
truly helped to spread literacy throughout European society to levels
never before known. Reading and writing were no longer the privilege
of the educated. Now merchants and tradesfolk could afford, at the
very least, their own copies of the Bible. In this sense, the technology
did change society. It certainly changed the Catholic Church—which,

before the first quarter of the sixteenth century was over, faced a Protestant reformation and challenges to its long-guarded authority and its monopoly on the interpretation of scriptures unlike any it had witnessed before.

In short, the spread of literacy threatened the clerical class of the Roman Catholic Church. There had been reformers during the Middle Ages, to be sure—and some were violently crushed by the church in collusion with the medieval state—but none possessed efficient and affordable means of distributing their tracts and their translations of the Bible that were later available to the reformers Martin Luther in Germany, John Calvin in Geneva, and William Tyndale in England.

To more fully appreciate the impact of paper, we need to trace its place in the history of writing itself. There's a long and winding path from the age of the Han dynasty, when paper was invented in China before the Common Era to Europe's first paper mills, the thunderous sound of their relentless trip-hammers frightening the residents of northern Italy in the thirteenth century. Before getting to paper, we must briefly discuss the media upon which humans depended for writing long before—in particular, clay, papyrus, and parchment.

The first system of human writing originated in Sumer, where some of the first cities formed out of the agricultural settlements around the lands of the Tigris and Euphrates Rivers in ancient Mesopotamia, what is now southern Iraq. Near the end of the fourth millennium BCE, cuneiform developed as an increasingly detailed means of marking trades and sales by intricate signs scratched on baked clay tablets.

According to Frederick Kilgour, the oldest ancient accounting records, maintained on soft clay tablets, used a variety of simple impressions: small, linear scratches laboriously carved into the clay to represent numbers and simple objects. Some of these tablets survived the ages and were uncovered in southern Iraq under the rubble of buildings that burned down before the soil of subsequent centuries covered them. But archaeologists agree that these tablets constitute simple accounts of the movement of goods in trade, with numbers marked in detail and then totaled, indicating that the earliest human writing grew from the needs of the ancient economy in Mesopotamia. This contrasts with the view of some historians that writing originated from the demands of ancient religious rituals or sacred writings.

Over time, Sumerian writing became more complex with the introduction of pictograms, small pictures (rather like a modern advertising logo) that represented an object or idea. "Sometime before 3100 BCE pictograms began to replace impressed signs; these pictograms were the beginnings of Sumerian script, the first written language," Kilgour writes.

> *The pictographic script, however, presented two problems: first it was difficult to write curvilinearly with a pointed stylus on wet clay; and second, there was no standardization of a single pictograph for a given object—at one time there was a cumulation of thirty-four pictographs for "sheep." Both difficulties were resolved by the invention of a triangular stylus that produced regular wedge-shaped impressions, various arrangements of which constructed uniform characters, known as cuneiform, from the Latin* cuneus, *meaning "wedge." The triangular stylus permitted a writing system that employed straight, rather than curvilinear, lines and encouraged the standardization of specific characters for specific words.*[2]

Cuneiform survived the culture that produced it, and by 2500 BCE, the new conquerors of Mesopotamia, the Akkadians, adapted cuneiform to represent the words and syllable sounds of their language. This spread among the other Semitic peoples of the Middle East until the eclipse of cuneiform tablets began in the second millennium BCE with the rise of phonetic alphabets, whose signs represented distinct sounds that made up the syllables of human speech.

This was no mean span of time, however distant it may seem to us. As historian Mark Kurlansky writes, "Clay tablets had the advantages of being cheap, readily available, and easy to write on. Their lack of portability, however, was an obvious drawback. Nonetheless, clay tablets were the world's primary writing material for three thousand years—a considerably longer period than the reign of paper up until now."[3]

It was the Phoenicians who invented the first alphabet in which twenty-two characters were assigned to represent the sounds that humans were capable of forming distinctly. It was quickly adopted and improved by the Greeks by 1100 BCE, when they introduced five new characters and changed four of the Phoenician consonants to vowels, making a total of five in their new system. By this time, the Greeks had discovered papyrus from the Egyptians, a more convenient material for writing curvilinear letters. By the sixth century BCE, papyrus began to replace clay tablets among the countries of the Near East.

Derived from the plant *Cyperus papyrus*, papyrus grew along the shallow banks of the Nile delta in southern Egypt. The reeds of the plant, growing from a triangular base, could reach sixteen feet in height with stalks up to two inches thick. The stalks were highly useful: they could be dried and bound to make rafts and boats and their inner layers were edible. But the Egyptians soon discovered a long-term utility: the stalks could be made into a durable writing material that could be rolled up for storage and travel.

Once chopped down, papyrus stalks were peeled, the thicker inner part of the plant yielding up to twenty inner layers. These were cut into strips and laid out side by side on a stone slab or table in a rectangular arrangement. Then another layer of strips was laid crosswise on top of these, and the entire sheet was pressed together, either manually or by hammering. The natural sticky sap exuded from the raw strips hardened so that the entire "mat" solidified into a single sheet, with some additional treatment.

In his *Natural History of Exotic Trees*, dating from the first century CE, the Roman writer Pliny the Elder described the rest of the process:

*The common paper paste is made of the finest flour of wheat mixed with boiling water, and some small drops of vinegar sprinkled in it: for the ordinary workman's paste, or gum, if employed for this purpose, will render the paper brittle. Those, however, who take the greatest pains, boil the crumb of leavened bread, and then strain off the water: by the adoption of this method the paper has the fewest seams caused by the paste that lies between, and is softer than the nap of linen even. All kinds of paste that are used for this purpose, ought not to be older or newer than one day. The paper is then thinned out with a mallet, after which a new layer of paste is placed upon it; then the creases which have formed are again pressed out, and it then undergoes the same process with the mallet as before. It is thus that we have memorials preserved in the ancient handwriting of Tiberius and Caius Gracchus, which I have seen in the possession of Pomponius Secundus, the poet, a very illustrious citizen, almost two hundred years since those characters were penned. As for the handwriting of Cicero, Augustus, and Virgil, we frequently see them at the present day.*[4]

The Egyptians wrote with a stylus made from reeds. Their ink consisted of crushed mineral powder mixed with water. One limitation of the new medium was that they could write on only one side of the sheet, and for lengthier texts, the sheets would have to be glued together into a long succession (some up to twenty feet long) and then wound onto one or two cylindrical sticks to make them more convenient for reading and for transportation.

The most famous surviving papyrus from antiquity is the *Egyptian Book of the Dead*. This is perhaps a misleading title for what was originally a collection of written spells and incantations intended for the use of those buried and waiting to enter the afterlife. The collection dates to the nineteenth century BCE, although scribes continued to add spells to it over the centuries. The *Papyrus of Ani*, which dates from 1250 BCE, is the most well-preserved copy. In 1888, it was stolen from the Egyptian government by a colorful and somewhat dodgy English Egyptologist and philologist named Ernest Wallis Budge.

For all of his shortcomings, Budge's own translation of the *Book of the Dead*, while dated stylistically, remains in print. The primary reason, as John Romer, one of his editors, points out in a recent edition, is that—considering that Budge was writing at a time when most of his contemporaries held ancient religion to be little more than savage superstition—Budge's translation took pains "to show us that a subtle living faith is held within these ancient texts. His commentaries describe a morality in the *Book of the Dead*, the discovery of which is usually credited to scholars of succeeding generations."[5] Ultimately, Budge was knighted for his opportunism, and the *Papyrus of Ani* now resides in the British Museum.

Because the papyrus plant was native to North Africa and the most abundant and sturdy specimens came from the Nile, the pharaohs were for centuries able to maintain a monopoly on the writing material, which they sold to the other nations around the Mediterranean. The Romans continued this monopoly after they conquered Egypt, and it lasted until the collapse of the empire. At this point, parchment—a thicker, more durable form of writing material made from animal skins—became the dominant means of record keeping in the West.

During the transition from papyrus to parchment, in the first centuries CE, a new medium emerged: wooden tablets played a small but crucial part in the transition from scrolls to bound books. And that was the invention of the codex, as it was originally employed—if not invented—by first-century Christians.

To make their texts easier to carry and consult, preachers of the Gospels assembled light tablets of wood and then bound them together so they could be written on both sides. As historian Harry Gamble writes, "The comparative evidence is instructive. Of the remains of Greek books that can be dated before the third century CE, more than 98 percent are rolls, whereas in the same period the surviving Christian books are almost all codices."[6] The early Christians, it seems, had an almost exclusive preference for the codex, and in this sense they were distinguished from the established bibliographic conventions of the era.

The codex was not thought of as a real book at the time it appeared. It was viewed as more of a simple notepad for private use only. The original Latin term, *caudex*, can mean a block of wood or a tree trunk. (It is interesting to note that the English word *book* derives from the German *Buch*, itself derived from the German *Buche*, which means beech tree or beechwood, from which early German books were made.) The custom was to link or bind several thin wooden shingles or writing boards together by string or leather cord, upon which the user could write when the need (or spirit) arose. The bound book as we have come to know it now is a direct descendant of this simple codex. In fact, the word *codex* continued to be used for books in the Middle Ages long after scribes ceased using wooden tablets for leaves.

According to Raymond Clemens and Timothy Graham, we know in more detail how wooden codices were made based upon a 1986 discovery in the Dakhleh oasis in Egypt.

*The excavations brought to light two codices, a farm account book and a compilation of Isocrates' Cyprian Orations. Both date from around 360 CE and exhibit similar construction, each having originally contained eight boards, or "leaves," on which writing had been entered. In each case, the entire codex was cut from a single block of wood. The individual leaves are mostly about 2.5mm thick, the outer leaves (which served as the covers) being a little thicker. When the artisan cut the leaves from the block, he made a mark on one side of the block (the side that would be the spine of the book) to ensure that when the book was finished, the leaves would maintain their original order.[7]*

Small books made from wood might have served the early Christians well in their travels. But by the time monks were making their own codices in the monasteries after the collapse of the Roman Empire, bookbinding had become not only an art but an industry based on an entirely different natural material—the skins of animals. Indeed, parchment became the "substrate upon which virtually all knowledge of the Middle Ages has been transmitted to us."[8]

As long as literacy was low and writing was the narrow province of clerics in monasteries and churches, from roughly the fourth to tenth centuries, parchment was an ideal medium for books and documents. Calves and sheep were the preferred source for skins in the north, but goatskin was also used, especially in Italy.

The skin on both sides of a hide could be used for parchment, but the inner side was easier to treat and also naturally smoother. The outer skin needed to be scraped carefully of any remaining hair or wool before it was soaked in a solution of water and lime for several days. After soaking, the outer skin would be scraped again until it was completely free of any hairs or roughness and then washed a second time in water. At this stage, the skins were mounted on wooden frames, stretched tautly with tension cords to prevent the skin from shrinking as it dried over the course of several more days. But more importantly for the end product, stretching brought about a change in the very structure of the skin, transforming it into a thinner material once fully dried.

The resulting parchment was both durable and pliable enough to be cut into sheets, which could then be folded into halves or quarters and bound between covers into books. The bookbinding could be as simple as a set of cords knotted between the covers. The covers were sometimes of tanned leather or smooth wood. As medieval artisans became more skilled, covers and bindings also became more elaborate. Scribes might draw small illustrations on the pages or stamp them with impressions made from miniature woodcut portraits dipped in ink, foreshadowing—though they did not know it—the coming of woodblock printing and movable type.

The problem, as the demand for parchment books rose, was the fact that parchment required a significant amount of hide. For example, it would take the skins of about three hundred calves to provide enough parchment to produce just one copy of the Bible. As the Middle Ages progressed into the eleventh and twelfth centuries, the population began to grow again and the economies of the European kingdoms prospered. The demand for parchment became too great to keep up with, given the laborious process of generating copies of books, let alone producing new ones. And this was true not just for books, but also parchment for book-keeping, contracts, and written laws.

At this stage, paper came to the rescue of a culture that was ready for it, just as it had for the culture that first invented paper more than a thousand years before: ancient China. As Kurlansky makes a point of reminding us:

*Studying the history of paper exposes a number of historical miscon-ceptions, the most important of which is this technological fallacy: the idea that technology changes society. It is exactly the reverse. Society develops technology to address the changes that are taking place within it. To use a simple example, in China in 250 BCE, Meng Tian invented a paintbrush made from camel hair. His invention did not suddenly inspire the Chinese people to start writing and painting,*

*or to develop calligraphy. Rather, Chinese society had already estab-*
*lished a system of writing but had a growing urge for more written*
*documents and more elaborate calligraphy. Their previous tool—a*
*stick dipped in ink—could not meet the rising demand. Meng Tian*
*found a device that made both writing and calligraphy faster and of*
*a far higher quality.*[9]

So, too, was the Chinese culture ready for paper when it emerged. Although today China officially reveres one man for the invention of paper in the year 105 CE, recent discoveries suggest that, like so many other inventions, paper was the end process of an evolution that took place over generations thanks to the efforts of many now-forgotten people.

"Writing about this key achievement three hundred years later, the official historian of the Han Dynasty, Fan Ye, declared that Cai Lun (until recently spelled Tsai Lun in the West) had 'initiated the idea of making paper from the bark of trees, hemp, old rags, and fishing nets,' and that once perfected, the process was 'in use everywhere.'"[10] Cai Lun was evidently a distinguished imperial administrator who specialized in manufacturing. However, when he came across the making of paper during his travels, he immediately recognized its potential.

But as Kurlansky writes, the legend of Cai Lun was undercut by the discovery of many ancient pieces of paper found during twentieth-century archaeological expeditions at sites that predate 105 CE. Most of these specimens were recovered from the deserts of central Asia, where the climate always has been dry. It's also possible that even older samples date from the second and first centuries BCE from the other parts of China, but because of the damp climate conditions, they did not survive.[11]

What the Chinese discovered then—and what remains true to this day—is that you need three key ingredients to make paper: water, cellulose, and a screen mold. The magic is in the chemistry among the three. Acquiring the cellulose is the hardest part, as it must be prepared by crushing natural vegetation and plant products into a pulp that can then be sieved and allowed to dry into sheets.

The first Chinese papers were made from the inner bark of trees, old fishnets, cloth scraps, and hemp gathered from frayed rope. These were soaked in water and then beaten to a fine pulp with a wooden mallet before being put back into water and stirred until the fine filaments could be seen floating in a film on top of the water. "Next, a scoop of the slurry was ladled evenly over a screen of coarsely woven cloth that had been stretched taut within a four-sided bamboo frame—what we in the United States call a mold, or mould in British parlance—and suspended between a brace of poles."[12]

In terms of the science, what was taking place in the vats was a process that chemists call hydrogen bonding, which allows free-floating cellulose fibers to attach to each other.

*What makes this process possible is the presence in cellulose of individual chemical units known as hydroxyl groups, meaning that many of the hydrogen atoms and oxygen atoms are paired together structurally in a way that permits them to act as single entities. When applied specifically to papermaking, some of the fiber-to-fiber hydrogen bonds take the additional step of replacing fiber-to-water hydrogen bonds as the pulp dries.*[13]

The quickest way to dry the pulp into sheets for paper was to sieve it in fine screens, or molds, and then let them dry.

Although the primary purpose of paper's rapid adoption in China was for writing—for which Cai Lun was revered—it's also true that the Chinese—and later the Koreans and especially the Japanese—turned paper to more beautiful uses in terms of artwork and interior design, from sliding doors with intricately designed screens of paper to spherical lanterns and the intricate detail of origami.

Paper's adopters in the West were perhaps less artistically inspired by the new material but were no less eager to exploit it. Paper reached Europe through an Islamic gateway in Spain as early as the eleventh century. Basbanes tells us that the first specific mention of a paper shop on the Iberian Peninsula dates from 1056,

*near the city of Xàtiva, southwest of Valencia, famous for its fine linens woven from locally grown flax. Precisely how the first papers were made there remains pure speculation, but given the analysis of surviving examples, we know that the fiber was rags and that the pulping . . . was probably done in stone troughs with water-driven pistons, or with trip-hammers known as stampers, though there is no definitive evidence to support either view.*[14]

But it was through Sicily, another Islamic toehold on the European continent, that the technique for papermaking worked its way into Italy and took an industrial root in the northern region near the Alps, where it set off an expansion throughout the rest of Europe. The first paper mill in Italy dates to 1235, possibly even earlier in the thirteenth century. In France one was recorded by 1348, in Austria by 1356, Germany by 1391, and England by 1494, a relative latecomer. (Scotland did not begin its own paper production until 1591.)

And so it came about that vertical water mills, which had long been employed to power trip-hammers for fulling cloth, powering forge bellows, and crushing ore, were now pulverizing rags to make the pulp for paper.

As it turned out, the industry arrived just in time for the printing press, which Johannes Gutenberg invented around 1450 (the exact date is not known) by borrowing the modern equivalent of a million dollars from a lawyer from a wealthy merchant family named Johann Fust.

Movable type already had been invented in China, but even before paper arrived, medieval artisans were using woodcuts to print playing cards and posters. They were also used to imprint the large letters and illustrations adorning the "illuminated" texts of parchment books in the libraries of the monasteries and fledgling universities.

As early as the twelfth or thirteenth century, both China and Korea may have been printing with movable type made from metal, which was an important improvement over wood and clay because imprinted figures on metal sheets did not wear out after a limited number of impressions, as inevitably happened with lighter materials. Nevertheless, wooden type was still being used in fourteenth-century China, as Kurlansky writes: "Wang Zhen, a magistrate and an agronomist who wrote voluminously, ordered the cutting of 60,000 characters in wood and used them to print his work as well as a local gazette. But with those characters, he could print only about a hundred copies.[15]

The drawback for Chinese printing was that it remained dependent on a system of thousands of characters, which made widespread dissemination of books impractical. For Islamic societies, in contrast, the whole idea of printing was dismissed as an unworthy pursuit in its own right. Certainly, printing copies of the Koran, let alone purely scholarly books, was out of the question, as Muslim religious and political leaders believed that their holy scriptures should always and only be copied by hand. Arabic letters were also more difficult to convert to movable type than the Roman alphabet, which had been adopted by the Europeans. As a result, the Europeans, with their increasingly mercantile sense of opportunity, exploited printing in a way that allowed them to catch up to both the Islamic and Chinese cultures in terms of book production and literacy. "This also meant that Europeans got to write history the way they wanted it to be read."[16]

Sometime between 1450 and 1455, Gutenberg produced a Bible in two volumes, thirteen hundred pages total, and he printed 180 copies of them. This may seem like a modest print run, but it took enormous resources and time, and he failed to sell most of them. In the end, Gutenberg failed to repay his loan to Fust, who then sued him and won possession of all of Gutenberg's printing presses and the remaining unsold bibles. Gutenberg languished and spent the rest of his career managing a print shop in Mainz, eventually dying in poverty. Two things are important to note: other Europeans were experimenting with movable type before Gutenberg. Also, there is no evidence of a technology transfer from Asia to Europe, and the fifteenth-century European invention of movable type is almost certainly an independent reinvention. Gutenberg's real achievement was not so much the invention of movable type, but the invention of the printing press and printing ink.

In the end, Fust set up his own shop, hired an enterprising former scribe named Peter Schöffer, and in effect became the first successful publisher of printed books, selling various titles to readers throughout Germany and France. By 1480, more than a hundred towns had their own presses, and by 1500, the number reached 326. More than thirty-five thousand editions of books had been printed in Europe by then.[17]

"The impact of printing on medieval society remains a debated topic among historians," Jeffrey R. Wigelsworth tells us.[18] Some scholars argue that the arrival of the new medium was revolutionary and transformed Europe. They suggest that the transition from a culture of scribes to one of printers and publishers was so momentous because of the amount of knowledge that could be preserved and proliferated by print simply dwarfed what had been preserved, until that time, by manuscript alone.

Other scholars have espoused a more evolutionary view, pointing out that the change to printed books merely continued—albeit much more rapidly—practices that were already common among handwritten book-makers. And hand-copied books written on parchment continued to be produced along with printed books for decades after the advent of print-ing. What is beyond question, however, is that printing and publishing, reading and writing, emerged for good from both cloister and university. Within only a few decades, Europeans were introduced to an intoxicating array of new kinds of books and tracts that they could read on their own.

## CHAPTER FIVE

# The Great Escapement

*This religious concern for punctuality may seem foolish to rationalists of the twentieth century, but it was no small matter to a monk of the Middle Ages. We know, for one thing, that time and the calendar were just about the only aspect of medieval science that moved ahead in this period. In every other domain, these centuries saw a drastic regression from the knowledge of the ancients, much of it lost, the rest preserved in manuscripts that no one consulted. Much of this knowledge was not recovered until reimported hundreds of years later via the Arabs and the Jews in Spain or, still later, from Byzantium. But time measurement was a subject of active inquiry even in the darkest of the so-called Dark Ages. One has only to compare Isidore of Seville's rudimentary notions of time in his* De Temporibus *(615 CE) with Bede's enormously popular textbook, the* Temporum Ratione *(725 CE)—written in the peripheral, tribal background that was Anglo-Saxon England—to realize the progress made in the field.*

—David S. Landes[1]

SOMETIME IN THE LAST FEW DECADES OF THE 1200S, MEDIEVAL craftsmen solved a long-standing mechanical problem that would revolutionize the technology of timekeeping. Who the first inventor or inventors were remains a mystery, though it was probably a small group of blacksmiths working with millwrights.[2] It's also likely that the escapement drive responsible for this innovation was first devised for something other than timekeeping: to automate the ringing of church bells.

Indeed, one clue to the first use of the escapement drive lies in the very root of the word *clock*, which did not come into common use until *after* the invention of the first mechanical timekeepers. It derives from the Old English *clok* (itself borrowed from the northern Latin word for bell, *clocca*), which referred to the striking of the bells in the church tower.

The makers of timekeeping machines until this period went by ponderous names such as *artifex horologiarum*: literally, maker of horologes, or hour logs. Of course, horologes classically referred to water clocks, a fact that has made it more difficult for scholars to determine which thirteenth-century records refer to purely mechanical clocks, since during the early decades of the escapement drive, the term was used for both types.

The inventors may have been in the employment of an influential bishop who was preoccupied with the building of a great church and who wanted to mount in its tower a more impressive machine to ring the church bells to signal the times of prayer during the day and evening hours throughout the year. But it may also be the case that an abbot simply wanted something more practical to be used as a timekeeping device for the monks in the halls of a monastery.[3] Within a very short time, however, it became obvious that the escapement mechanism they developed was useful for more than just ringing bells.

It could also be used to power a gear train, which could drive multiple hands on the face of machines designed to track the movements of the sun, the moon, and even the planets all the year round. The great clock in the Abbey of St. Albans in Hertfordshire, England, completed twenty years after the death of its designer, Abbott Richard of Wallingford in 1336, was just such a machine.

Although they did not realize it, the inventors of the first great medieval clocks were following in the footsteps of the Greek engineers who designed the Antikythera mechanism to predict the motions of the sun, moon, and planets back in the first century BCE. The abbot of Wallingford in particular came to grips with many of the same challenges that may have confronted Archimedes (if indeed, as seems likely, he was the original inventor), in delicately coordinating the gears to drive multiple models of the heavenly bodies.[4]

Destroyed during the dissolution of the monasteries under King Henry VIII, the St. Alban's clock was not destined to endure, but other simpler timekeeping machines did. Because cathedral building was a competitive activity among cities during the twelfth and thirteenth centuries, adding a mechanical timekeeper with the latest gearing technology must have appealed to many a bishop. The first recorded mentions of mechanical clocks between 1280 and the early 1300s indeed show most were constructed for mounting in churches in cities such as London, Oxford, and Canterbury and on the continent in Milan and Cambrai.

The challenges presented by the machine that drove the first mechanical timekeeper were not too far from the craftsmen's everyday experience. Blacksmiths and millwrights were the engineers of the age, men with practical knowledge of the underlying forces that made their mills and forge bellows work. They knew little of what moderns would call principles of engineering. They had no word for "force" as we understand it, but they knew it when they saw it. They knew not only how to use their hands and tools to build important laborsaving devices like water mills, camshafts, and trip-hammers, but they also knew how to drive these using the energy supplied by water and wind. Now, having harnessed the energy of water and wind, it was perhaps an obvious next step to try and determine how to harness the energy of gravity. But that was a much more challenging problem, one that apparently had never been solved before.

Sun and water had been the standard means of timekeeping until the Middle Ages. Because the twelve hours of the day were traditionally counted from sunrise to sunset, and the twelve hours of the night from sunset to sunrise, this convention led to unequal hours, hours that varied in length with the season of the year, with long daylight hours during the summer months and long nighttime hours during the winter months. Time was curiously relativistic in a sense, hours shrinking and elongating along with the seasons.[5] But time was not an abstraction for people in the ancient world and the medieval world. It was identified with the duration of concrete processes like the irrigation of fields or the period between religious observances, duly measured in the path of the sun's shadow or the flame's consumption of a candle during the night—or the flow of water out of a graduated bowl.

The Greeks and Romans relied on water clocks and sundials, the descendants of great obelisks that the Egyptians built in the desert sun of the pharaohs, close to the equator. The hours of the day were inscribed in a great perimeter about their base. But as can be imagined, obelisks and sundials were of little practical use for much of the year in northern Europe, where the shadows (when the sun was not obscured by clouds) grew long and the daylight fleeting in the autumn and winter.

Water clocks were more reliable. The Greeks produced the most efficient, known as clepsydras (literally "water thieves") around 325 BCE. These were small, consisting of an empty bowl with a precise hole at its base. This punctured bowl floated within a larger basin filled with water. Slowly, as the punctured bowl took on more water through the hole in its base, the water level gradually reached markers inscribed at different points on the inside of the bowl as it filled. Such a water clock could be emptied and reused repeatedly; for example, to mark the duration of an irrigation cycle on a farm. More elaborate water clocks relied on a small outflowing trickle of water from a large basin to fill a separate hollow cylinder in which a floating marker slowly rose, pointing out the hour on a measuring rod attached to the inside of the cylinder.

Arab inventors, perhaps taking their cue from earlier Chinese models, built more elaborate water clocks—not only to signify the hour, but to delight travelers from other lands. They began to employ gear chains powered by waterwheels to set automatons in motion to emerge, like robots, from small doors in the walls of their timekeepers to announce the hour and track the motions of the heavenly bodies. Instructions for the construction of Arab water clocks survive in many texts, such as *The Book of Knowledge of Ingenious Mechanical Devices* by the Turkish engineer al-Jazari, written between 1204 and 1206. Again, their purpose was as much to enchant as to signal the hour, and in this regard, the water clocks, though uncommon, no doubt made an impression on European travelers who brought tales of them back home in Europe. Notable among these was a great clock presented by the Sultan of Damascus, al-Ashraf, to Emperor Frederick II Hohenstaufen in 1232. Described by one source as a "marvelously fashioned tent," it featured astronomical simulations, including the motions of the sun and stars and the hours of the day and night.[6]

To improve the design of water clocks, Spanish inventors in the late thirteenth century, working for the Christian King Alfonso X of Castile, drew up elaborate plans for a clock using mercury instead of water to improve the resistance necessary to control the weights operating the gears of the clock. Unfortunately, there is no evidence that the machine was ever actually built.[7]

But in the north, as the age of barbarian invasions waned and civilization spread slowly once again, monks used calibrated candles to mark the passage of hours between prayer times. The flame slowly burning past each hour marked on the side of the candle indicated when it was time to commence prayers. But candles had to be replaced each day, they consumed a lot of wax, and they were not reliable in the drafty halls of monasteries and churches where the slightest breeze might blow them out. Water clocks like the ones from Islamic Spain soon spread to the more affluent monasteries and churches in Europe, and the more elaborate ones were employed to ring the "alarms" for prayer.

Although it's logical to assume that the weight-driven clock was inspired by a practical need for something better than a water-driven device, something that (in theory) required less maintenance, historians note that the first weight-driven clocks were not a practical replacement for their water-based relatives in the short term. Maintenance was more expensive, requiring the services of an experienced blacksmith if anything went wrong with the gear trains or weights.

There was another source of inspiration: astronomy. From a technical perspective, the drive to produce a purely mechanical clock seems to have been driven by two ingenious astronomical tools of Greek origin: the astrolabe and the equatorium. In terms of design, they certainly inspired the great dials that would become the faces of every church clock in Europe.

The astrolabe, invented by the Greeks, was adopted and improved by Arab astronomers and navigators to track the fixed stars and determine prayer times, no matter how far the user was from Mecca. In its simplest form, it was no more complicated than a slide rule set within nested dials that diagrammed the positions of the brightest stars and constellations of the zodiac. But by the very end of the tenth century, Persian astronomers—most notably al-Biruni—had added additional insets in order to track the movements of the sun and the moon to more accurately determine daily prayer times as well as the dates for Ramadan, the month of fasting.

A simpler cousin of the astrolabe, the equatorium was designed to track the movements of the planets only. Europeans came into possession of both tools through Spain when Christian armies began to reclaim

Astrolabe. IMAGE COURTESY OF ADAM JARED APT

territory from the Muslim incursions of the centuries before. But to clockmakers, both devices recommended themselves as the perfect "face" for a new kind of clock.

Abbott Richard of Wallingford, one of the greatest astronomers of his age, was certainly inspired by the astrolabe for the building of the unique clock at St. Alban's. By the time he was appointed abbot in 1327, the escape drive to automate it completely by weights had already become known to clockmakers. Wallingford wanted to combine the utilities of the astrolabe and the equatorium to follow the progress not just of the day and night, but also of the sun, the moon, and, according to one account, perhaps even the progress of the known planets during the course of the entire year.

To track these heavenly bodies as well as the hours of the day, many geared wheels would need to be coordinated to move the "hand" of the clock as well as the picturesque representations of the sun and planets in their proper places around one great dial on the face of his machine. This included the moon's monthly track around the Earth and the sun's plodding track through the zodiac throughout the 365 days of the year. On top of this, the clock was also expected to activate the striking of the church bells on the hour. The question was how to make this all work without a dozen or so monks doomed to spend their entire lives turning great gear chains in the bell tower.

One drive would suffice, and fortunately for the abbot, someone had already invented it. But it must have taken a great deal of trial and error. One astronomer from the time who offers us a glimpse of the nature of the challenge posed by the weight-driven mechanical clock was Robert the Englishman, who taught for a while at the University of Montpellier in France. In one of his commentaries from 1271 discussing the work of a fellow English astronomer known as John of Holywood (Sacrobosco), Robert wrote the following:

> *Clockmakers are trying to make a wheel which will make one complete revolution for every one [hour] of the equinoctial circle, but they cannot quite perfect their work. . . . The method of making such a clock would be this, that a man make a disc of uniform weight in every part*

*so far as could possibly be done. Then a lead weight should be hung from the axis of that wheel, so that it would complete one revolution from sunrise to sunrise, minus as much time as about one degree rises according to an approximately correct estimate.*[8]

The main challenge facing the inventors was how to use weights to "program" the wheel to make one complete revolution in the space of an hour: to turn just enough to mark the time without simply spinning out of control until the weights plummeted to the ground. That required a braking mechanism.

Such a braking mechanism became known as the verge and foliot. As you can see from the illustration, the device consisted of a rod with two protruding pallets at right angles, separated along the rod's length, and a crossbar at the top bearing adjustable weights. Its job was to manage the rotation of the escape wheel.

The escape wheel was a disk of uniform weight mounted on an axle designed to drive the progress of one hand on the clock's dial. For simplicity, let's say the hour hand. This axle's rotation was driven by a weight wound with a cord. Without any interruption, the weight would descend under gravity to the floor and the axle would spin freely, too rapidly to be of any use in timekeeping. The verge and foliot's job was to brake the weight's acceleration into discrete one-second progressions. In order to do this, the escape wheel needed to fit with the verge and foliot perfectly. Note in the illustration that the disk of the escape wheel is not smooth around the edges. It bears spokes (or as in the illustration on the next page, saw teeth, like a king's crown turned on its side) that extend the same length and are separated along the perimeter of the wheel at equal distances.

The pallets of the verge were of equally precise design. The verge is the long, thin vertical rod on which the foliot is mounted. The distance between the verge's two pallets, one high and one low, had to match the height, or diameter, of the weight-driven crown wheel exactly. If this sounds like a familiar device, that's because it is: the verge is a kind of camshaft, an axle with two precisely placed cams (pallets) to control reciprocating motion—in this case, pausing the escape wheel in its revolutions. The pallets on the verge must point in two different directions, 90 degrees away

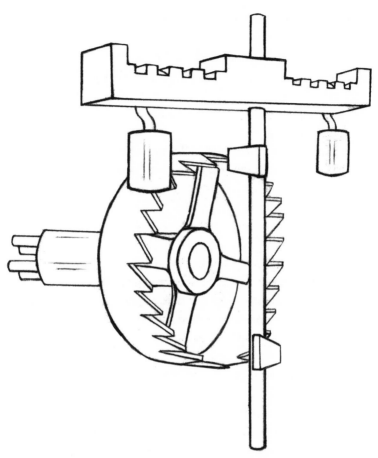

Clock with verge and foliot. ILLUSTRATION BY RYAN BIRMINGHAM

from each other. The pallets also must be positioned to lock with the top and bottom of the escape wheel spokes simultaneously. The weight-driven crown wheel requires evenly distributed spokes, each of which protrudes from the perimeter just far enough to interact with the pallets of the verge. The foliot consists of a crossbar with two balance weights on either end. The balance weights are adjustable, so they could slide inward or outward along the horizontal foliot, depending on how much inertial resistance is desired. But they always must be equidistant in order to maintain consistent resistance to the force of the escape wheel pushing against the pallets.

73

In action the escape drive of the mechanical clock works thus: The descending weight turns the escape wheel on its axle. This causes the top spoke of the wheel to strike against the high pallet on the verge, pushing the pallet away at a right angle. In the same motion, the bottom spoke of the wheel strikes the lower pallet. The net effect between the two is a consistent interference in the escape wheel's rotation by both upper and lower pallet on the verge so that the wheel rotates—but slowly, thanks to the resistance provided to the pallet by the weights on the foliot. In this manner, the clock receives a dependable amount of power from the weight in order to tick smoothly and for a long time. The first clock probably required hours, if not weeks, of trial and error to determine the ideal weights and sizes, both for the crown wheel and for the verge, with a primary goal being to require rewinding only once a day and perhaps, as its improvement progressed, even less frequently.

As noted earlier, the escapement mechanism may have been designed at first to automate the ringing of church bells at designated times, but it quickly evolved. The great clock designed by Richard of Wallingford, who was trained as an astronomer (and as the son of a blacksmith had a working knowledge of how to make gears), depended on a different escapement, one that Wallingford perhaps believed was superior. Wallingford utilized a variation of the drive, which he wrote about in the early fourteenth century, but the underlying principle was the same—an alternating brake on the gravitational energy used to rotate the main axle driving the clock.

In 1327 in his *Tractatus Horologii Astronomici*, Wallingford wrote that his clock's drive, which he called a *strob* escapement, relied on two escape wheels mounted on the same axle, so that their radial teeth, or pins, were staggered between them. A semicircular brake was mounted on a verge-like rod with adjustable weights between the wheels; this would turn back and forth (like the pallets on the verge and foliot design), with each passing pin generating the same beat-by-beat progress for the timekeeper. There is some question as to whether this approach actually came first, before the verge and foliot, but the latter design became more common over the next century.

The success of the new machine quickly spread, leading to the production of mechanical clocks throughout Europe. These new clocks inhabited monasteries and the towers of churches, where the hours could be announced by the ringing of bells set in motion each hour. The "stroke" of midnight or noon is still a term used to signify the moment when the hands of the clock meet at twelve. At that instant, the great clock activated an arm to strike the bell in the church, letting the whole town know what time it was—and what chores attended the new day.

Very soon, meaning within a few decades, mechanical timekeepers began to dominate medieval culture. The hour was not the only solid ticking standard, set by sixty minutes. Soon, as the design of clocks continued to improve, each minute was in turn set by sixty seconds.

Churches began to ring bells automatically on the hour, then, not long after, on the half hour and then the quarter hour. Time began to mean something more than the duration of the daylight hours alone; it no longer flowed like St. Augustine's subjective stream. Time became discrete, more concrete, something to be measured in little units.

The day no longer needed to be measured in terms of the duration from sunrise to sunset as it was in ancient times. The French seem to have been the first to use clocks to institutionalize the twenty-four-hour day of equal hours independent of the seasons. Although it took a few centuries, this eventually became standard in Europe.

Time, in a sense, became the first quantum revolution back in the Middle Ages. In its wake, life would never be the same again (although Augustine's haunting description of time still inspires physicists). The consequences were far reaching. As early as 1316, for example, Dante incorporated clocks into canto 10 of *Paradiso*, the third part of his epic poem, *The Divine Comedy*, suggesting the lovely sound of tinkling bells struck on the hour by heavenly clocks. But not all references to clocks would be inspiring. In 1595, a little more than three hundred years after the first clock began to tick, William Shakespeare gave voice to the more ominous implications of time in its new mechanical guise:

> *I wasted time, and now doth time waste me;*
> *For now hath time made me his numbering clock:*
> *My thoughts are minutes; and with sighs they jar*
> *Their watches on unto mine eyes, the outward watch,*

*Whereto my finger, like a dial's point,*
*Is pointing still, in cleansing them from tears.*
*Now sir, the sound that tells what hour it is*
*Are clamorous groans, which strike upon my heart,*
*Which is the bell: so sighs and tears and groans*
*Show minutes, times, and hours: but my time*
*Runs posting on in Bolingbroke's proud joy,*
*While I stand fooling here, his Jack o' the clock.*

Shakespeare's *Richard the Second* was based on the short life of an English monarch who lived less than a century after the first clocks began to appear. It's an interesting twist perhaps that he was also the grandson of King Edward III, the monarch who rose to power at the very time that Richard of Wallingford was building the first great clock of written record at St. Alban's.

In Shakespeare's imagination the dethroned King Richard came to learn painfully how the time he took for granted and wasted on his throne now haunted him in the silence of prison, where he could count the seconds like a mechanized clock but was entirely helpless to do anything about his situation. Time, for Richard, became a slow death.

The tragic melancholy that Shakespeare attached to mechanical timekeeping likely would never have occurred to the bishops and towns-folk of the fourteenth century. It became a matter of civic pride for a city or town to have its own great clock. By the mid-1300s, most of the largest cities on the European continent had at least one great clock in the cathedral or the most prominent church. St. Gothard's in Milan, for example, was the first clock to strike equal hours on the hour in 1335. The public clock in Padua followed suit by 1344, Florence's in 1354, Bologna's in 1356, and Ferrara's in 1362. By 1370, clocks proliferated to the point that there were disputes about synchronizing the time. King Charles V of France ordered all churches in Paris to ring on the hour and quarter hour according to the time set for the new clock built for him by Henri de Vick in the Palais Royal that same year.

Although the clock started as an important tool for the church's liturgical needs, it soon became crucial for the success of tradesmen and the growing merchant class. According to one twentieth-century historian, "The bells of the clock tower almost defined urban existence. Timekeeping passed into timesaving and time-accounting and time rationing. As this took place, Eternity ceased gradually to serve as the measure and focus of human actions."[9] For example, an architect who could more accurately gauge the hours spent by his masons, carpenters, bricklayers, and blacksmiths could more efficiently project the costs of building a new cathedral or bridge. Mechanical time became the standard for financial prosperity.

But the change was not overnight, and more recent scholars suggest caution in assessing how quickly public clocks transformed the social attitudes of Europeans. There is strong evidence that sundials and local solar time remained the principal means of time control for many communities until the nineteenth century. Certainly there was no great change in people's sense of time before the second half of the fourteenth century. Public mentions of clocks prior to this in France and Italy are few and often vague as to whether the clocks were mechanical. As one historian writes, "Old habits die hard. The clock on the tower of the Palazzo Vecchio at Florence was repaired in 1512, and not until then was it altered so as to indicate the hours in our present way, twelve before and twelve after midday. The 'new' way was described at the time as the French style."[10]

Timekeeping, as we understand it now, is truly an invention of the Middle Ages. There are some in our own busy day and age who believe we have become the clock's slave. But without the mechanical clock and the precision of timekeeping it bestowed on civilization, there would never have been a Scientific Revolution.

# CHAPTER SIX

# The Cathedral Crusade

*In the twelfth century, masons working on the church of the monastery of Obazine in France threw down their tools in anger one morning and hurled abuse at the Abbot because he had thrown away a pig they had killed and which they had been looking forward to eating. The Abbot was a vegetarian!*

—JEAN GIMPEL[1]

AS WE SAW IN THE LAST CHAPTER, THE FIRST MECHANICAL CLOCKS were mounted on the towers of churches and cathedrals where they could ring out the hours for all the town and the countryside to hear. Clocks and church towers would go hand in hand throughout the centuries, as if they were designed for each other. In one sense they were. The greatest and tallest church towers were built just decades before the first clocks were ready to be mounted. One might say that the twelfth and thirteenth centuries were the building boom period of the Middle Ages, at least in terms of churches. But the instigators weren't real estate developers as we know them. They were abbots and bishops of the church, kings and lords, and the leaders of the ever-growing towns and cities throughout Europe.

The ancient Romans set a high standard for stonecutting and architecture—an area in which they indisputably outshone the Greeks. The remains of aqueducts, bridges, and pagan temples throughout the lands of Rome's empire still draw tourists to marvel at what they were able to accomplish with stone and concrete, the latter an invention that predated the Romans but that they perfected.

According to T. Roger Smith, "The Roman Empire had introduced into Europe something like a universal architecture, so that the buildings of any Roman colony bore a strong resemblance to those of every other colony and of the metropolis; varying, of course, in extent and magnificence, but not much in design. The architecture of the Dark Ages in Western Europe exhibited, so far as is known, the same general similarity. Down to the eleventh century the buildings erected (almost exclusively churches and monastic buildings) were not large or rich, and were heavy in appearance and simple in construction. Their arches were all semicircular."[2]

But the stonecutters and masons of the eleventh century caught up to their ancient forbears quickly. At first content to mimic the designs of the buildings of the past when it came to building new churches (and new meeting halls for kings or manors), they followed the simple rectangular layout of the Roman basilica, which featured colonnades on either side of the building, an apse at one end, and a semicircular dome erected over the center.

The builders of the Middle Ages quickly found themselves eager to experiment, albeit within familiar constraints of the ancient building styles at first. Over the course of the late tenth and early eleventh centuries, their approach to building design revealed a definite shift away from thinking in terms of flat, undifferentiated planes of walls and ceilings, which was the standard in the basilica. The Roman style relied on thick stone walls to support domed roofs. As a consequence, however, windows were few and small, and the interiors of buildings were dark even during the day. Medieval builders grew restless with this template, and though they did not have the theoretical knowledge to discard the classic design and start again from scratch, necessity forced them to rethink how they could reshape the flat, rectangular interiors of the churches to form new dimensions within the old. In short, they found themselves moving "toward discovering means of delineating the spatial units and volumes contained within buildings; indeed, it hardly overstates things to say that the articulation of spatial volumes was central to the architecture of the late eleventh and early twelfth centuries."[3]

Moreover, this shift in design involved rethinking the use of the interior rectangular space to accommodate new forms of miniature construction; for example, to allow the statues of saints to be mounted in small niches, wall enclosures, and not only on the ground floor but in higher spaces near the ceiling. The ceiling itself could be supported by something more dynamic than heavy stone arches. All of this meant coming to grips with design problems that were much more complex even than those the builders of ancient Rome confronted. In tackling these challenges, the master builders of the Middle Ages would, during the twelfth and thirteenth centuries, distinguish themselves to the degree of initiating a whole new discipline: that of the architect. By the eleventh century, they were beginning to build castles, meeting halls, manors, bridges, and, most striking of all, entirely new churches that continue to amaze visitors to the cities of Europe: the great Gothic cathedrals.

The skills and the aspiration to build churches and cathedrals initially arose to a significant degree from the experience of building castles. That said, at the outset of the era of cathedrals, medieval stone castles themselves were a relatively recent development in Europe. Until the era of the Viking invasions finally subsided in the tenth century, most European fortresses were hastily built from timber and vulnerable to fire. It was in the late tenth and early eleventh centuries that European kings found the time to muster the resources to build castles of stone and mortar and to raise high towers.

To economize, builders quarried what stone they could from as close to the site of a castle as possible in order to cut down the time and cost of transporting the stones over long distances. If they were lucky, they could also mine the limestone they needed to make mortar. Mortar was produced in kilns on-site; the limestone was heated to high temperatures to the point that it reduced to a white powder. It was then mixed with sand and water to produce the quick-drying mortar cement for the masons to layer between stones. The stones for building castles and later cathedrals and churches became known as ashlar, because they were cut as uniformly as possible, almost like bricks, with the same dimensions, so that they could be more evenly laid out for the foundations and walls.

Some of the castles that have survived this early era of medieval building still show signs of how they were constructed, as Wigelsworth tells us. Holes throughout the castle walls reveal where scaffolding was built during their construction. "This archeological evidence reveals that temporary structures were erected to serve as working platforms and for material storage as the castle wall took shape and became higher. Holes in the stones of towers indicate the use of spiral scaffolding."[4]

The master builder would have relied on a team of masons and blacksmiths on-site to make the hammers, augers, chisels, and trowels the builders needed, including the plumb bob and the level, which the master builder used to ensure the walls were straight and stones aligned.

Corfe Castle in Dorset on the south coast of England is one of the oldest stone castles still standing. It was built soon after William the Conqueror succeeded in his invasion of England, and its remains—still impressive—are open to the public. Colchester Castle, another keep also built on the orders of King William, was designed by one of his bishops, Gundulf of Rochester, and built on the foundations of a Roman temple between 1069 and 1169 CE, which was constructed by recycling much of the original building material. Ten years seems to have been the average amount of time for castle construction, although during the crusades

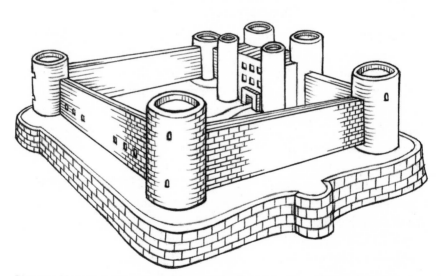

Diagram of stone outer walls and towers of a feudal castle. ILLUSTRATION BY RYAN BIRMINGHAM

in the Middle East, they were built more quickly due to the constant threat of attack.

It's interesting that we see in Colchester the direct involvement of a bishop in the building of castles for the king. This was not to be a rare partnership, and it would inspire one French abbot to an entirely new approach to building within a few short decades of the launch of the crusades at the end of the eleventh century.

In 1095, Pope Urban called for the first holy war to retake Jerusalem. Although he did not realize it at the time, he launched a period of almost three centuries of largely unsuccessful invasions of the Holy Lands by European kings. The first attempt would succeed but at a huge price in human suffering, as the knights who journeyed to retake Jerusalem from its Muslim conquerors just as brutally seized and pillaged Christian and Jewish settlements along the way. As Jonathan Riley Smith writes in his history of the Crusades, in the buildup to mustering the armies of liberation,

> *There were no means available for screening recruits for suitability, other than the decisions of the magnates on the composition of their households; indeed there could not have been, because . . . as pilgrimages crusades had to be open to all, even psychopaths. The appetites of the violent may well have been sharpened by disorientation, fear and stress as they sacked their way to the East. The vicious persecution of Jews in France and Germany, which opened the march of some of the armies, was marked by looting and extortion and the passage of the crusaders through the Balkans was punctuated by outbreaks of pillaging.*[5]

The campaign for the First Crusade inadvertently touched off a different kind of crusade at home, however. Although many clergy accompanied their lords on the journey to Jerusalem, their bishops had a different problem on their hands: how to inspire the increasingly productive townsfolk who were left behind.

> *So a new kind of movement was needed in which the townsmen could participate, and such a movement was launched in the middle of the twelfth century. It happened about the time when the king of France*

*was absent on the second crusade, and when the regent ruling in his
stead was Suger, abbot of the great monastery of St. Denis near Paris.*[6]

When it came to building a new church, Suger was more than a man
with a plan. He had a vision of bringing more light into the buildings
dedicated to his god. With his first project, the rebuilding of his abbey
church, which began in 1137, the age of the cathedral really began.

At the time that Pope Urban issued his call for the First Crusade,
Suger was a boy of fourteen and already devoted to a life of prayer at his
monastery. A childhood friend of King Louis VII of France, he grew up
to become one of the king's most trusted advisers and, as noted earlier,
he ruled France as regent in the king's absence during the second crusade. There's some irony in the fact that, a few decades later, a similar
close friendship between the English King Henry II and his highest
ranking church official, Thomas Becket, the archbishop of Canterbury,
would sour and lead to Becket's notorious murder by some of the king's
closest henchmen.

When it came to his new construction, the religious inspiration for
Suger was tied deeply to the writings of the saint who lay buried in the
crypt of his abbey church: St. Denis. Mistakenly believed at one time to
have been the Greek judge from Athens who was converted to Christianity by St. Paul (and referred to in the Acts of the Apostles as Dionysius
the Areopagite), in fact St. Denis was a third-century martyr killed in
France by the Romans during the persecution of Emperor Decius sometime after 250 CE. He became the patron saint of France and the church
where he was buried evolved into the monastery that became the center
of the town north of Paris that still bears his name.

Abbot Suger wanted to call together craftsmen from every walk of
life to rebuild his abbey's church in honor of St. Denis. He took full com-
mand of the project, and the church was built between 1137 and 1148,
consciously employing some of the new structural innovations that he
had learned masons were already employing in churches as far away as
the cathedral at Durham in England.

But he went further, making notes on how he believed he could
work some of his patron saint's theological reflections on the nature of

creation into the design of his church. As it turns out, there was, over the centuries, quite a case of mistaken identity as to the authorship of the *Theologica Mystica*, which made such an impression on Suger as he thought about his abbey church's new design. Had he been aware of the mix-up, it's unlikely it would have changed his determination. In this sense, the entire rebuilding of his church, as George Duby writes, was an exercise in applied theology:

> *Naturally enough, that theology was based on the writings of the abbey's patron saint, Denis—that is, Dionysius the Areopagite, or so it was believed. For the remains of the kings of France lay near the original tomb at this site, that of Dionysius, the Christian martyr of the region called France. Not only Suger but also all of his monks and all the abbots who preceded him identified this hero of the conversion to Christianity with St. Paul's disciple, Dionysius the Areopagite, traditionally held to be the author of the most imposing mystical construct in the history of Christian thought. The text of that work, written in Greek during the very early Middle Ages by an unknown thinker, was preserved in the monastery of France, that is, Saint-Denis. In 785, the pope had given a manuscript of the work to Pepin the Short, king of the Franks, who had been a pupil at Saint-Denis. In 807, another copy was sent to Louis the Pious, Emperor of the West, by Michael, Emperor of Constantinople. The first Latin translation of it, a poor one, was done by Hilduin, an abbot of Saint-Denis. Then, during the reign of Charles the Bald, Johannes Scotus Erigena, who had a better command of Greek, produced a far better translation with a commentary. So it was that the* Theologica Mystica *was held in awe at Saint-Denis. On it Suger's thinking and his art were based.*[7]

For all of this book's influence, however, it should also be noted that Suger traveled extensively throughout France and saw examples of the designs he later used.

The older abbey church at St. Denis had been built in the Romanesque style of the basilica, in which the roof of the church was low and little room was spared for windows. The Romanesque style was, in a general sense, a transitional style between the ancient Roman architectural style and the later Gothic style, although the former did not simply fade out with the latter's ascendance. Romanesque buildings, usually churches, featured square exterior towers with round arches and a great round vault, or barrel vault, built over the central part of the church, forming the large signature dome that could be seen from afar. Adding to the rectangular plan of the ancient Roman style, the Romanesque churches also projected two transepts on either side so that the entire church formed the shape of a cross. The apse was always on the eastern end of the church, facing the dawn, and the entrance was opposite on the western side of the building. Romanesque was itself an evolution from the more classic basilica church from the early Middle Ages, which featured a rectangular building and flat roof.

As a result, the building was gloomy even by day and more so with candlelight. Suger wanted to infuse his new church with daylight filtered through stained glass windows, another recent innovation of medieval glassworkers. Here we need to lay out what Suger incorporated into his

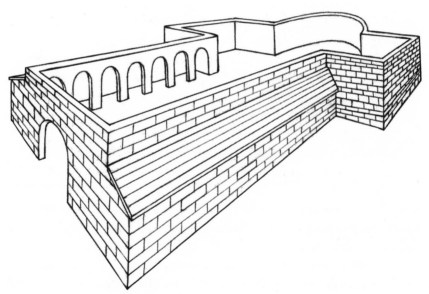

Basic layout of the medieval Romanesque basilica. ILLUSTRATION BY RYAN BIRMINGHAM

abbey church design: it was deliberate and it was about maximizing the use of light. Fortunately for posterity, he left many of his thoughts in writing.

*Here, Suger's reflections on his innovations at the abbey church of St. Denis—the archetypal, and effectively the prototypical Gothic church—are extremely important. These comments stress the spiritual and religious significance of a new way of building, which would increase the light, especially through stained glass (this certainly made the church's interior more jewel-like, but not necessarily any lighter). The idea that architecture could transform a church into a miniature heaven on earth, a foretaste of the celestial Jerusalem, was a prime stimulus.*[8]

What was also important to Suger was impressing the pilgrims and visitors to his church. The new innovations of his builders offered an opportunity to do just that. As historian Jon Cannon writes,

*The process by which Romanesque transformed into Gothic can be seen as a kind of fifty-year exploration of the aesthetic, engineering and geometrical potentialities of the pointed arch. One of the roots of this transformation is the Norman discovery of the rib vault: as this new feature became de rigueur, its implications began to lead in a new direction. Romanesque was thus transformed from within, even as it reached its maturity.*[9]

Three key distinguishing characteristics of the Gothic cathedrals were pointed arches, rib vaults, and flying buttresses. In contrast to the rounded arches of the ancient Roman and later Romanesque designs, medieval builders began to "point" the arches in more of a triangular shape in order to ensure that the weight of the ceiling and roof above was directed directly down. Beyond a limited height (and weight), rounded arches distributed its burden outward to the sides rather than downward, eventually driving more weight against the walls of the building.

Rib vaults are a series of small, rounded arches shaped from limestone, that, together, are strong enough to support the construction of a stone roof above the cathedral ceiling. Until the mortar holding the

Comparison of round Romanesque and pointed Gothic arches. ILLUSTRATION BY RYAN BIRMINGHAM

stones of the roof together dried, the rib vaults were the main support. Once the stone roof solidified with the mortar, the rib vaults served as decoration. But like pointed arches, they contributed to the main goal: allowing the use of windows that could be installed much higher in the overall structure of the church than before and inviting more light into the building.[10]

Even more crucial to the cathedral's structural integrity was the employment of flying buttresses, which were designed to support thinner stone walls from the outside of the cathedral—much thinner and lighter walls than those of the basilica style. Again, in service to the grand design of allowing more daylight into the church, flying buttresses also allowed greater space for stained glass windows as well as the addition of side balconies, or "colonnades" with their own arched ceilings and columns, all above the ground floor where pilgrims gathered in the nave, the central part of the church.

The highest vaults were built for the cathedrals of Notre-Dame de Paris and those in Chartres, Rouen, and Reims. The Notre-Dame and

The rib vault provided additional support to vaulted ceilings and stone roofs. ILLUS-
TRATION BY RYAN BIRMINGHAM

The flying buttress provided external support for thinner walls and larger windows.
ILLUSTRATION BY RYAN BIRMINGHAM

Chartres cathedrals were about equal from the floor of the nave to the highest point of their vaulted ceilings—110 feet. The vaults at the cathedrals of Rouen and Reims, Pacey tells us, "were about 125 feet above the floor in both cases, while Amiens Cathedral, begun in 1220, had a vault soaring up to 140 feet above the floor."[11] But this invited a response from the citizens of Beauvais, another town of northern France, which, like Amiens, was a big commercial center. The rebuilding of the cathedral in Beauvais began in 1247, with plans to raise its vaults to 157 feet. Very soon the bishop and citizens of Cologne decided to match them.

These undertakings were often decades long, and in some cases neither the bishop nor his master builder lived long enough to see the cathedrals completed. Work continued year-round, except during the winter months when it was too cold for mortar. According to Jean Gimpel's account of the building of Westminster Abbey, at the height of the summer season, between mid-June and late July, Westminster's bishop employed between 390 and 430 men: masons for carving statues; masons for the stonecutting; carpenters to build the scaffolding as well as the frames, under the master builder's direction, into which the cathedral walls and arches were assembled stone by stone; sanders for smoothing the stonework; blacksmiths for making the tools; glassmakers for stained glass; roofers; and general laborers.[12]

"Between 1150 and 1280," Pacey writes, "about eighty cathedrals were built or rebuilt in France, many in the cloth towns in the north of the country. In effect, Suger had inaugurated a new sort of crusade, one that was enthusiastically pursued by merchants and townsmen in the way in which the military crusades had previously engaged the enthusiasm of knights and nobles. Thus one historian thinks it appropriate to talk about this wave of church building as 'the cathedral crusade.'"

The cathedrals were the most ambitious and expensive projects undertaken during the Middle Ages. Although there is no way to calculate the cost in terms of modern currency, it is not a stretch to liken the decades-long undertakings as equivalent to the multiyear highway-building projects of today. Building a cathedral was far more expensive than building the most elaborate castle. Indeed, when the great age of building had begun to wind down around the end of the thirteenth century, historians estimate the cost of a cathedral would have been four or five times as much as large castles at that time, around £15,000.

Bishops were not shy about how they raised funding. Kings and dukes could be relied on for some gifts, but the majority of the money came from the people who lived in the towns where the cathedrals were built. Local merchants, guilds, and trades groups could be counted on to contribute— often in return for the granting of indulgences, in which the bishop gave them "time off" from purgatory in the afterlife, a tradition that scandalized some church and monastic leaders and later became a source of contention between the leaders of the Protestant Reformation and the popes. But the bishops also raised money from the town's poorest inhabitants as well as the pilgrims drawn to the city to see the regular, highly publicized exhibitions of saints' relics (most of dubious authenticity to modern eyes).

The practice was not to the liking of one of the most influential leaders of the monastic movements. St. Bernard of Clairvaux famously criticized the bishops and abbots of other orders for raising money via the display of relics. The eyes of the people, he wrote, "are feasted with relics cased in gold, and their purse strings are loosed. They are shown a most comely image of some saint, whom they think all the more saintly that he is the more gaudily painted. . . . The church is resplendent in her walls, beggarly in her poor; she clothes her stones with gold, and leaves her sons naked."[13]

Over the course of 150 years, more churches and cathedrals would be built throughout Europe, until the late thirteenth century, when the fervor began to decline after the collapse of the vault of Beauvais Cathedral in 1284. Beauvais had been the highest vault in the world. After it caved in, enthusiasm for building began to wane.

Scholars disagree about the cause of the decline of cathedral building. According to one scholar, "After 1280, there was a notable decline in northern French tall-building activity. Although this has often been ascribed to a collapse, due to a hidden design flaw, of the original (quadripartite) high vaulting of Beauvais Cathedral that occurred in 1284, it was mainly due to a significant downturn in the local economy, and soon thereafter also by the outbreak of the Hundred Years' War between England and France."[14]

But even before the outbreak of war, others claim that builders had lost their enthusiasm for competing for the highest vaults and spires. The faith that animated the cathedral crusade was dying and there were signs of fear and reaction throughout the rest of medieval society. With more exaggeration than scholars today would condone, Jean Gimpel writes, "Roger Bacon, who wrote an amazing treatise on educational reform which would have revivified medieval Christianity, was thrown into prison and died around 1292. Freedom of expression, which held a place of honor in universities, was stifled. Canon law clashed with Roman law, which was being revived, and nationalism put in an appearance. The Pope's authority and prestige declined. The great monastic orders founded no new abbeys and began to have difficulty in recruiting lay brothers."[15]

To make matters more ominous, the long period of mild climate that helped launch the growth of agriculture and thus the economic expansion of the eleventh through the thirteenth centuries came to an end. Some of the cathedrals would be abandoned and fall into ruin even before the arrival of the Black Death in the middle of the fourteenth century.

Today, thousands of tourists and modern pilgrims visit the cathedrals at Notre Dame, Cologne, Reims, and the many others that have survived the ages. In contrast to the first pilgrims, they come from everywhere around the world, by plane, by ship, by train, or by automobile. But none of them comes on foot as the first pilgrims did. For them, it was a much more arduous journey, one with a spiritual urgency that modern travelers may not share. It was also likely the journey of a lifetime—the only "vacation" from their ordinary lives that they were likely to have, the opportunity of so many people to leave their homes and travel on foot to see a great cathedral and perhaps the relics of patron saints.

There is some irony in the fact that although the bishops and abbots—the great cathedral builders—wanted to create a destination for travelers and a natural locus around which their towns and cities could gather, little thought was given to improving the network of roads and highways that would facilitate the trip.

The people of the Middle Ages not only failed to improve upon the craft of the ancient Roman road builders, but the improvement of the roads seems never to have occurred to them. As Lorenzo Quilici writes,

*The roads that the civilization of Rome created throughout the ancient world represented a political event of universal significance. It is sufficient to note that these structures still today often constitute the basis for the modern road system, not just in Italy and the countries around the Mediterranean, but also throughout Europe and the Middle East. Only the invention of modern paving techniques and the construction of limited-access highways resulted in additions to the ancient system.*[16]

Indeed, using medieval England as one example, the responsibility for roadbuilding and highway maintenance seems to have fallen on the shoulders of the property owners (often monasteries and manors) who lived closest to the roads. But, using another example, in the port city of Valencia in Spain, basic and quite onerous duties were, by custom, assigned to the *frontalers*, whose property fronted on public roads and lands. Only very late in the Middle Ages did civic authorities and kings set aside funds and make edicts regarding the maintenance and construction of better roads and bridges.

In spite of this handicap, we know that millions of people throughout the Middle Ages traveled—and often traveled far—on foot, and not only for pilgrimages to the cathedrals that were becoming legendary. There was another component.

It is perhaps difficult to appreciate the effect that the long dirt and gravel roads of the medieval countryside had on the people who lived closest to them, especially now when many of us own a car or can access public transportation. Modern people in the developed world do not need to walk very far to any destination anymore. It's hard to see anything romantic about interstate highways (although some American poets and writers have created something of their own romance around the long roads of the American Midwest).

In his book, *Inventing the Middle Ages*, Norman F. Cantor touched on this aspect of medieval social life, which was uniquely captured in the novels of a fellow medievalist, J. R. R. Tolkien, perhaps best known for his Lord of the Rings book series. Although not a fan of the fantastical aspect of the stories that made them worldwide bestsellers, Cantor noted how Tolkien dramatized the circumstances and conditions of long journeys undertaken by ordinary people on foot without escort or the convenience of the horses and carriages that the aristocratic classes could depend upon.

> *This kind of distant journeying by obscure people over long distances, for one reason or another, we know from stray references, was a much more common occurrence at all times in the Middles Ages, but especially after 1100, than we might a priori predict from the kind of primitive transportation system the medievals had access to. People of modest social status in surprising numbers traveled long distances, mostly on foot. This is a strange fact of medieval life, and [*The Lord of the Rings] is centered on this event. Tolkien convinces us that the way this happened was that some local village leader got it into his head that he had to do something to help or save his people, something had to be carried a very long distance, some contact vaguely imagined had to be made, and off the person and two or three companions went on their incredibly long, footsore journey. These journeys were rarely documented for us in the Middle Ages and almost never in detail. Tolkien, by imagining such a journey, has graphically re-created an important but poorly understood facet of medieval social life.*[17]

As we see in the next chapter, an entire intellectual revolution would be launched by monks and scholars journeying on foot to distant lands in order to find better source texts of astronomy and philosophy. They would discover much more than they bargained for.

# From Greek to Arabic and Back Again

*It was indeed for renewal of learning in the Latin world that the translators had worked in the late eleventh and twelfth centuries. Their aim was progressively fulfilled. If individual amateurs quickly welcomed the translations, appreciation of the versions of Greek and Arabic science and philosophy developed more slowly in the organized studia and universities. In the course of the thirteenth century, however, the program of studies came to rely largely on the fruit of the translators' labor.*

—Marie-Thérèse d'Alverny[1]

THIS BOOK IS MAINLY ABOUT INNOVATIONS IN MEDIEVAL TECHNOLOGY. But along with progress in food production, clothing, and manufacturing paper, there was also an intellectual awakening in the twelfth and thirteenth centuries that would have far-reaching consequences for the later development of science and the Renaissance.

Even here, technology played a crucial role. The intellectual revolution of the later Middle Ages was triggered in no small part by the demand for better technology. Or more precisely, by the demand for better *instructions* about how to use technology, especially the tools of astronomy. These tools included carefully written tables with precise recordings of celestial movements and books of planetary theory, the most dominant of which was the great work of Claudius Ptolemy, *The Almagest.*

If you were a monk in France or Italy, intent on projecting the most accurate dates of each year that Easter and other holy days should fall, the only way to get your hands on the most reliable copies of these texts meant finding a good translation from Arabic or Greek. And the only way to acquire these was to hit the road.

Such was the origin of the medieval "translation movement," an enterprise that largely took place in Spain, Sicily, and southern Italy between the years 1130 and 1275. What started out as a search for better translations of astronomical and astrological guides led first to the discovery of Arabic treatises on the astrolabe and the astronomical tables of the great Persian mathematician and astronomer al-Khwarizmi, whose name transfers into English as *al-goritmi*, and ultimately evolved into the term, algorithm, and who worked in Baghdad in the ninth century CE.

But this soon led to the discovery of other Arabic textbooks on medicine, philosophy, and theology.[2] As Arab philosophers and doctors had long adopted the ancient works of Aristotle, medieval scholars from Europe found themselves rediscovering source texts of the Greek philosopher that for centuries they had known only from references in other works that summarized his thoughts, often not accurately.

The wanderers who came to Spain and Sicily from Italy, France, and the British Isles did not know it at the time, but with the rediscovery of Aristotle's works and the voluminous commentaries written about him by many Arab and Jewish philosophers and scientists, Christian scholars would face a revolution—indeed, what some worried bishops and theologians would deem to be an information overload.

For several hundred years—from the fourth to the ninth centuries—the study of philosophy and science in Europe had stagnated as its widely separated cities and ports struggled to adapt to wave upon wave of migrations, both internal and external.

This cultural stagnation accompanied a decline in the political and military power centered in Rome. In 285, the emperor Diocletian divided Rome's vast territories into western and eastern administrative regions. A little more than a hundred years later, in the decades after the emperor Constantine's death, the empire split in two unequal parts, leaving one ruler governing the larger eastern portion (the Byzantine Empire) from

Constantinople (present-day Istanbul), and a less exalted ruler governing the smaller western portion from Rome. The main preoccupation of Rome's rulers, emperors in name only, became defending the Italian peninsula from constant invasion.

The split between east and west isolated the two cultures that had long nourished the empire in tandem: the Greek culture of the Hellenistic world including Greece itself, Asia Minor, Egypt, Syria, Palestine, and farther east into Persia and the Latin culture of Rome, acting as the chief means of communication between the regions of Gaul, Spain, Italy, the British Isles, and North Africa. As a result, the number of scholars fluent in Greek dwindled in the west. Roman educators became increasingly reliant on outdated sourcebooks.

The disparate group of scholars known to history as the encyclopedists did their best to collect what knowledge remained after the disintegration of the Roman Empire. But the facts they accumulated were not only derivative, they were often garbled in transmission from one successive generation to the next. The gaps in their sources and knowledge did not stop the encyclopedists from citing the works and opinions of the great philosophers and scientists, such as Plato and Aristotle, Archimedes and Euclid, as if they themselves had direct knowledge of the original texts.

There were bright spots. Anicius Manlius Severinus Boethius (480–525), Roman politician, scholar, and author of *Consolation of Philosophy*, mastered Greek at a young age and had ambitious plans to translate the majority of Aristotle's works into Latin. Unfortunately, he managed to complete translations of only a few of the philosopher's shorter works on logic before his professional career demanded his complete attention. The western empire's military leader in Rome, Theodoric the Ostrogoth, held Boethius in such high regard that he made the scholar master of the palace officials as well as the head of his civil service.

Such favor was bound to make Boethius enemies, and it led to his downfall. The story is complicated, and historians continue to debate whether Boethius was guilty of the accusations brought against him by others in Theodoric's court. Theodoric was a fair-minded ruler of Italy and the European provinces still under Roman control. But Boethius found himself caught between his duty to the Roman leader and his own desire to help maintain the unity of the entire empire under the emperor Anastasius in Constantinople.

Theodoric was a Christian like Boethius. But he was also an Arian at a time when that sect of Christians, which did not accept the traditional dogma that God the father and Jesus the son were one, were being persecuted by Greek Christian authorities. For this and other reasons, Theodoric wished to establish more independence from Constantinople, and the enemies of Boethius found an opportunity to set him up— principally, it seems, by producing evidence to Theodoric that Boethius had at one time defended the persecution of Arians.

This led to his imprisonment for a time, during which Boethius conceived and wrote his philosophical masterpiece. If he had any friends in the Roman Senate, they were unable to stop a majority from condemning him. Boethius was ultimately sentenced to torture and executed in Pavia in 524. With his death, the West lost a direct link to its Greek scientific and philosophical heritage from which it would not recover for another six hundred years.

As the Roman Empire disintegrated, the migrations and invasions of peoples across the continent—especially the movement of Germanic populations—prevented the establishment of any lasting order. The Goths, the Lombards, and the Franks all moved into central and southern Europe from the northeast during the first four centuries of the Common Era. In the fifth century came the more destructive invasions of the Huns, and finally in the eighth and ninth centuries Viking raids wiped out whole settlements and monasteries from Ireland to as far south as Spain. All of these depredations, along with the shift from urban back to isolated, rural communities on the continent, worked against the establishment of any robust centers of learning.

During this time, the monasteries became the only safeguards of what remained of scholarship in the West. The preservation of science was not directly a concern of St. Benedict in the sixth century or of what was to become the order that bears his name, the Benedictines. But because it was necessary to preserve and pass on the sacred texts of the Christian faith—the church's theological tradition—monastic orders like the Benedictine monks and, centuries later, the Cistercians, the Carthusians, and others also preserved and copied as much of the pagan tradition as they could.

The Benedictines, who placed a great emphasis on stability in the daily life of prayer and meditation, copied and preserved much of the art, poetry, and philosophy of the pagan world that they considered valuable to a life of contemplation and prayer. (Indeed, Monte Cassino, the monastery that Benedict founded south of Rome in 529, would attract some of the most important translators of Greek science into Latin in the tenth and eleventh centuries.)

But the monastic orders soon found there were practical reasons for the preservation of ancient knowledge and philosophy. In England, for example, the Venerable Bede (672–735), a Benedictine and historian of England working in the late seventh and early eighth centuries, compiled what astronomical knowledge he could find for the purpose of drawing up more accurate calendars. Like Pliny and Isidore of Seville before him, he wrote his own version of *De Rerum Natura* (On the Nature of Things), largely concerned with astronomy. Although his text uncritically repeated some bogus knowledge from the derivative collections of older centuries, one of his later works, *De Temporibus* (On the Divisions of Time), added a store of firsthand material, including accurate assessments of the tides based on data from different ports and the calculation of the Easter calendar for the years 532 through 1063, the mathematical discipline called computus.

An official proclamation calling for a recovery of learning had to await the coming of Charlemagne (742–814), who restored, however briefly, a political unity to the European continent not seen since the days of the Roman Empire. In recognition of this, as well as his preservation of the papacy from the incursions of the Lombards, Pope Leo III crowned him the first Holy Roman Emperor on Christmas Day in 800.

Charlemagne ordered cathedrals and monasteries to found new schools dedicated to educating a new generation of scholars. He himself learned to read Latin, although he never achieved any facility in writing it. His ability to speak several languages was sharper: he could not only speak Latin, he also learned some Greek and was conversant in many of the dialects and languages of the regions he ruled.

Charlemagne's aspirations for a new renaissance were quite modest in the long run, destined as they were largely "to peter out in the political confusion and massive invasions of the bloody, terrifying ninth century."[3] It was not until the relative stability of the tenth century that kings and popes made a truly successful call for education reform.

## GERBERT OF AURILLAC

In 967, Count Borrell of Barcelona visited a Benedictine monastery at Aurillac, in the central part of France. The abbot of the monastery introduced him to one of his brighter pupils, a boy of peasant background named Gerbert, and asked the count if he would take Gerbert back to Barcelona. There Gerbert would enjoy access to a broader selection of scientific texts than were to be found anywhere else at the time.[4]

Young Gerbert went to the Spanish March (modern Catalonia) for two years of study at the Cathedral School of Vic. When he joined Count Borrell on his trip to Rome in 969, he had some understanding of the counting board and the astrolabe and was able to use Arabic numerals in place of Roman numerals, which were still dominant in Europe, to make easier calculations.

Gerbert was the beneficiary of what appear to be the earliest translations of Arabic scientific texts on mathematics and astronomy, which were completed in Spain in the 900s. The young scholar would enjoy the friendship of the Holy Roman Emperor, Otto II, whose son he tutored. Otto III would eventually repay his teacher by one day appointing him to succeed Gregory V as pope. Gerbert took the name of Sylvester as pope, but he was not the pontiff for long (four years), due to the political maneuvering between the young emperor and the rebellious people of Rome.

Gerbert fled the eternal city before he died from malaria in 1003, just one year after the luckless young Otto had died from it, but his wider legacy as one of the first European intellectuals schooled in Arab science has endured.

Of particular note is that for seventeen years (972–989) Gerbert was a teacher at the Cathedral School at Rheims. He was considered an inspiring instructor—and it has been said that he passed on to his pupils what he himself had learned of Arabic science and the astrolabe in writing—but recent scholarship has questioned how proficient he actually was. In the tenth century, a monk named Lupitus of Barcelona produced a copy of a "how-to" astrolabe manual called *Sententiae Astrolabi*, which was a translation from the Arabic, perhaps done by Lupitus himself, although it is not known for certain. According to the contents, the book seems to be related to al-Khwarizmi's treatise on the astrolabe. We know there was a copy of *Sententiae* in the Ripoll Scriptorium when Gerbert was there. And in his *Geometria*, Gerbert describes how to use an astrolabe to make certain calculations, but there is no definite evidence that he actually was using the instrument himself rather than working from an illustration of one. More recently, scholars also have questioned whether Gerbert in fact learned any Arabic astronomy during his Spanish travels.[5]

What is more certain is that he handcrafted a spherical model of the heavens as it was understood from the geocentric Ptolemaic system of the time. Gerbert's armillary sphere, as it became known, simulated the motions of the stars and planets and employed wires attached to the sphere to map the constellations. Among his enthusiastic pupils were future scholars Fulbert of Chartres, Adalberon of Laon, and John of Auxerre, who would go on to found some of the leading cathedral schools of the next century.

## Adelard of Bath

Another cleric who followed in Gerbert's footsteps was Adelard of Bath (1080–1152). Adelard's family seems to have come over to England from Lorraine, but as one historian wrote, he "comes out of the shadows in the 1120s in the West of England," where he worked as a cleric and published his most significant translations and treatises.[6]

With Adelard, we see some of the first real inspiration of European scholars to go out on the road to acquire and translate the technological treatises of their acknowledged superiors. As a young man, by his own account, Adelard journeyed across the continent to Italy and Sicily. It is possible he also traveled as far east as Syria, but there is no clear evidence apart from allusions in his writings.

Adelard made his reputation not only as a translator, but also as an enthusiastic proponent of Arab science, from which he and likeminded scholars such as William of Conches (tutor to the future King Henry II of England) developed a new naturalism, or what they called the doctrine of secondary causes. What may sound like a dry theological debating point actually became an important principle for later medieval scientists (or "natural philosophers" as they were called at the time). Without taking anything away from God as the ultimate cause of all, the new doctrine of William and Adelard's generation of scholars emphasized reliance on reason to study the natural causes of physical phenomena without resorting to miracles to explain them.

Although William was the more enthusiastic teacher, Adelard was the more articulate defender of this school of thought. In *Questiones Naturales*, which he addressed to his nephew Nepos, he wrote: "The natural order does not exist confusedly and without rational arrangement, and human reason should be listened to concerning those things it treats of. But when it completely fails, then the matter should be referred to God. Therefore, since we have not yet completely lost the use of our minds, let us return to reason."[7] There's a certain bite to that last sentence—an impatience that suggests a testy wit that still reverberates throughout the centuries.

Adelard was at one point in his travels drawn to Salerno, a medical center south of Monte Cassino where Arabic texts were also being translated into Latin. The most notable of these were translated by a North African convert to Christianity known as Constantine the African, who came to Monte Cassino in the middle of the eleventh century and translated the medical works of Hippocrates (ca. 460–370 BCE) and Galen (ca. 129–200 BCE).

While in Salerno, Adelard may have learned to read and write Arabic himself. There is no clear evidence that he personally translated Arab scientific texts for use in his surviving treatises. It seems more likely that Adelard befriended Arab scholars in Sicily and Italy and then transliterated what they conveyed to him into his own work.[8]

In 1126 or shortly thereafter, Adelard completed Latin versions of al-Khwarizmi's astronomical tables. These had been revised and updated for the rulers of the city of Cordoba before they were adopted by Adelard. Once his Latin translation was complete, it was possible for European astronomers to more precisely plot the positions of the sun, the moon, and the five known planets (Mercury, Venus, Mars, Jupiter, and Saturn), making it easier to cast accurate horoscopes.

Astrology would come under attack by some religious authorities in the generations after Adelard, but during the time that he was writing, the field of study—in spite of its occult associations—was considered a legitimate science and was later defended by both Albertus Magnus and Thomas Aquinas. As Adelard told his nephew in his *Questiones Naturales*, "Of course God rules the universe, but we may and should enquire into the natural world. *The Arabs teach us that.*"

Adelard was writing now at the leading edge of what would become a concentrated wave of translation activity. Whereas European scholars would rediscover the great works of classic literature during the Renaissance, the twelfth-century scholars were hungry for science and philosophy. What they transmitted to the schools and universities in France, Italy, Germany, and England laid the foundations for a European science that quickly overtook the technologies and science in the East. One center of this activity was in Spain.

## The Emergence of a Translation Movement in Spain

On the southern Iberian Peninsula, the Umayyad caliphate in Cordoba collapsed in 1031 after almost three centuries of rule. Al-Andalus, as it was known, splintered into smaller kingdoms. While the Berbers of North Africa, followers of a more rigid Almohad tradition of Islam, invaded to restore some unity to the land, the far northern cities like Santiago de Compostela in the region of Galicia in the northwest corner of the peninsula were vulnerable. Christian kings eager to expand their domains beyond northern León-Castile and Aragon-Navarre began to retake them.

As news of the reconquests spread, the great monastery at Cluny, France, began to encourage spiritual pilgrimages to the cathedral of St. James in Santiago de Compostela once it was firmly under the control of the Christian kings. People from all walks of life from all over Europe sought to purge their sins by making their way to the great church and doing penance.

Pilgrims soon became a familiar sight on the roads over the Pyrenees to Santiago de Compostela. The cathedral is, according to legend, the final resting place of James the Greater, one of the Sons of Thunder (the sons of Zebedee) and one of the twelve apostles of Jesus. According to the Acts of the Apostles, James was beheaded in Jerusalem in 44. Tradition holds that his body was brought back to Galicia, where he had first preached the gospel.

Well in advance of Pope Urban II's call to crusade in 1095, the abbots of Cluny were, in the words of one recent historian, "also annexing to that thought of pilgrimage for penance another new and potent idea. St. James had become the symbol of the fight-back of Christians in Spain against Islamic power."[9] And the Cluniacs' investment in the pilgrimage routes to Compostela would influence the balance of power between Christianity and Islam in Spain.

The recapture of the bells of Santiago de Compostela planted the seed for wider ambitions, not only in the minds of Spanish kings on the Iberian Peninsula, but in the minds of the bishops of Rome. It gave impetus to push eastward all the way to the other side of the Mediterranean where Jerusalem—the Holy City—lay under Muslim rule.

When Byzantine Emperor Alexius I sent ambassadors to Pope Urban II to muster armies to defend Constantinople against the Turks, Urban had already been informed of the success of the Spanish princes as they reclaimed Toledo in 1085. If the Muslims could be driven out of Spain, there was good reason to believe that they could also be driven out of the Holy Land. This thinking would prove faulty in the long run. The Crusades not only failed to liberate Jerusalem for long, but they led to horrific massacres of the very people they were intended to free.

By this time, the philosophers and theologians in the growing number of cathedral schools on the continent had received reports of the astronomical charts, medical books, and treatises waiting for them in the cities the Spanish kings were recapturing. Indeed, the translation movement in Spain essentially began in Zaragoza, the Muslim kingdom in the northeast of the peninsula, after it was retaken by the Aragonese in 1118. The victory made available a wealth of scientific manuscripts in the palace library. The so-called Ebro Valley translators who set to work translating them into Latin included scholars such as Hermann of Carinthia, Robert of Ketten (in England), and Hugh of Santalla, working independently on astronomy, alchemy, astrology, geomancy, and, in Herman's case, also the Koran. Within a generation, the same manuscripts were available in Toledo.

In contrast to how the crusaders would treat the residents of Jerusalem after taking that city in 1099, the Christian dukes leading the reconquest of the Iberian Peninsula were far more humane. King Alfonso VI of Castile, for example, captured Toledo without bloodshed. The residents were allowed to remain in the city and to keep all their possessions. True, many of the Islamic upper classes departed the city and many of those remaining converted to Christianity. But Alfonso described himself as a king "of two religions," and by and large he did not upset the centuries of tolerance that had existed among Muslims, Christians, and Jews under the Umayyad caliphate based in Cordoba. The majority of the city's residents spoke both Arabic and a Romance dialect and were able to live in relative peace (although, predictably, as elsewhere in Europe, Toledo saw periodic pogroms against the Jewish minority).

The city had been a leading center of scientific learning in the Muslim culture of al-Andalus. Although the new archbishop was indifferent to scientific matters, the cathedral at Toledo was to become the major center of translation activity for the next century.

## GERARD OF CREMONA

Indeed, it was not long after Adelard finished his translation of al-Khwarizmi's astronomical tables—perhaps as early as the late 1140s—that a young Italian scholar named Gerard of Cremona (ca. 1114–1187) came to Toledo in search of Ptolemy's classic compendium of Greek astronomy, *The Almagest*, which he had been informed was available there in Arabic. Gerard was not able to find a Greek version in his home country, though history has subsequently shown that a new translation was made from the Greek at about this time by Henricius Aristippus in the Norman kingdom of Sicily (see page 112).

Gerard became a canon in the cathedral at Toledo and spent the rest of his life there, working on translations of at least seventy texts from Arabic to Latin. Gerard usually did not affix his own name to the translations he made, but his students compiled a list not long after his death. They published the list along with a generous eulogy in his last work, a translation of Galen's medical treatise, *The Art of Medicine*:

*Although from his very cradle he had been educated in the lap of phi-losophy and had arrived at the knowledge of each part of it according to the study of the Latins (*Latinorum studium*), nevertheless, because of his love for the* Almagest, *which he did not find at all amongst the Latins, he made his way to Toledo, where, seeing an abundance of books in Arabic on every subject (*facultas*) and, pitying the poverty he had experienced among the Latins concerning these subjects, out of his desire to translate, he thoroughly learnt the Arabic language, and in this way, trustworthy in each . . . he read through the writings of the Arabs (*scriptura Arabica*), from which he did not cease until the end of his life to transmit to Latinity, as if to a beloved heir, in as plain and intelligible way as was possible for him, books of many subjects whatever he esteemed as the most choice.*[10]

Gerard studied Arabic but not deeply enough to translate directly himself. He depended on an Arab Christian named Galippus. Others who came to Toledo also found it more practical to call upon the expertise of Christian and Jewish residents already fluent in Arabic to translate from Arabic to the Spanish dialect of Toledo, which they could then render into Latin for the benefit of scholars back home in the European schools.

It is curious that all of the great figures behind the translation movement in Toledo were foreign born. Even the bishops of Toledo of the time were of Frankish birth. None was native to the city. This fact underscores the reality that the translation movement was primarily an export business. It was for the scholars on the other side of the Pyrenees that Toledo's translators labored so long and so hard.

At first glance, there seemed to be no real rhyme or reason in how the Toledo translators approached the documents they translated; they "groped somewhat blindly in the mass of new matter suddenly disclosed to them. Brief works were often taken first because they were brief and the fundamental treatises were long and difficult; commentators were often preferred to the subject of the commentary."[11]

By the end of Gerard's life, he had translated Ptolemy's *Almagest*; Euclid's *Elements*; Aristotle's *Posterior Analytics*, *Physics*, *On the Heavens*, *On Generation and Corruption*, and *Meteorology*, as well as the major works of mathematics and astronomy of the great Muslim scientists al-Kindi and al-Farabi, as well as Ibn Sina's *Canon of Medicine* and al-Khwarizmi's *Algebra*, to name just a handful.

Indeed there may have been some deliberate program of attack between Gerard and another canon at the cathedral, his colleague Dominicus Gundissalinus. Gerard tackled the mathematical and astronomical texts of both Greek and Arab origin, while Gundissalinus, working with the Jewish scholar Avendauth, concentrated on the philosophical books and commentaries of Ibn Sina, the Persian physician and philosopher who became known to the Western world by his Latinized name, Avicenna, and al-Ghazali, his fiercest critic and opponent of philosophy in general. According to one recent account:

> there was a clear division between the translating activities of Gerard of Cremona and Dominicus Gundissalinus. The one favored the authentic works of the Greeks and their Arabic commentators, the other favored Avicenna's philosophical approach to philosophy and the reading-matter of contemporary Jewish scholars. Considering that both Gerard and Gundissalinus worked within the precincts of Toledo cathedral, it is hard to believe that they were unaware of each other's work, or inimical to it.[12]

John of Seville characterized the translation of Avicenna's *De Anima* thus: "Ibn Dawud took the text and pronounced the [Arabic] one word at a time as they were spoken by the people (vulgariter), while Gundisalvo [Gundissalinus] wrote down the Latin equivalent to each of these words as he heard them."[13]

## IBN RUSHD

It is ironic that the man European philosophers and theologians would regard as one of the most influential Arab commentators and philosophers, Ibn Rushd (1126–1198), was writing at the time that Gerard and Gundissalinus were translating the texts of his predecessors in medicine and science. Ibn Rushd was a native of Cordoba, in the south of Spain, and his work covered a broad range of topics in medicine and philosophy, but it would not be translated until the generation of scholars after Gerard.

Known to Thomas Aquinas in the next century as Averroes, Ibn Rushd was—thanks to the "Angelic Doctor"—destined to have a much greater impact on the European mind-set than he ever did on his fellow Muslims. Indeed, not long after Gerard of Cremona died in 1187, Ibn Rushd found his own works under attack by the Islamic religious leaders of Cordoba.

Born in 1126, Ibn Rushd came from a prominent family of judges in that city. He was educated as a physician but felt strongly drawn to the study of pure philosophy. A fortuitous meeting with the caliph of Cordoba, Abu Ya'qub Yusuf, led to a full-time position for the young doctor. Yusuf also enjoyed the study of philosophy in spite of the more conservative attitude of the Almohad dynasty, and he employed Ibn Rushd to write new commentaries on Aristotle to clarify all the major works. Many previously had been translated into Arabic from earlier Syriac versions, but Yusuf found them difficult to understand.

This period, the last quarter of the twelfth century, was the time during which Ibn Rushd wrote his most influential treatises, adopting a rationalist point of view—not unlike the view we have seen propounded by Adelard and William of Conches—that would profoundly influence the pursuit of science in the later Middle Ages.

After the death of Abu Ya'qub Yusuf, the caliph's son al-Mansur continued to support Ibn Rushd for a few years, but by the 1190s, more conservative members of the Islamic schools began to attack the philosopher, and due to some questionable charges made against him, Ibn Rushd was, at the age of sixty-nine, banished from Cordoba to the nearby village of Lucena. Though he did not suffer the fate of Boethius, his books were banned; indeed, most were eventually burned—though not all. He was allowed to return to Cordoba one year before he died in 1199 when he was seventy-three. But he was not destined to have anything like the influence on Muslim thought that Ibn Sina and al-Ghazali had.

Ibn Rushd was never to know the honor and esteem in which his works would be held by his country's enemies to the north of the Pyrenees. When Aquinas wrote his *Summa Theologica* during the latter half of the thirteenth century, he referred to Ibn Rushd as "*The* Commentator." When it came to any discussion of Aristotle, Aquinas rarely failed to reference Ibn Rushd's assessment of any question—even when he disagreed with it.

## THE SICILIAN CONNECTION

Historians know less of the other clerics in Toledo working alongside Gerard and Gundissalinus. Many are scholars whom we know only by their names on the manuscripts, by references to their work in translation made by their contemporaries—or from the generation that followed.

Not all the translators of the Toledo "school" were foreigners like Gerard. John of Seville, who appears to have emerged from the community of Arab converts to Christianity (known as Mozarabs), grew up in what became the Kingdom of Portugal. Of the aforementioned Ebro valley translators who were a generation before Gerard, Hugh of Santalla was another homegrown translator. His contemporary Robert of Ketten came from England like his countryman Adelard years before and Plato of Tivoli, who distinguished himself in Barcelona by translating astrological texts from Arabic to Latin with the aid of a Jewish scholar, Abraham bar Hiyya, came from Italy.

John of Seville's main interest was in translating astrological texts of Arab astronomers such as al-Farghani and Masha'allah. Hugh translated astrological texts attributed to Ptolemy, as well as treatises that would strike modern readers as bizarre, including one by Masha'allah on the art of divining the future from the shoulder blades of animals. Note this sequence of gradual integration of Arabic science in the Christian West: the earliest Aristotelians in the European schools were more interested in astrology than philosophy. That changed as more Arabic science and philosophy were adopted.[14]

Robert of Ketten worked with Herman of Carinthia to translate a copy of the Koran for Peter the Venerable, Abbot of Cluny. On his own, he translated al-Khwarizmi's *Algebra* and al-Kindi's *De iudiciis astrorum* (Concerning the Judgments of the Stars). Plato of Tivoli translated works of Archimedes and Ptolemy.

By this time, translators were hard at work in the courts of the Christian kings of Sicily who had retaken the island from Muslims in 1091. Although less programmatic than their fellow scholars in Spain, the translators working in the twelfth century, first for the court of the Norman King Roger II (1095–1154) and later his son William I (1131–1166), made equally important translations of Greek scientific and philosophical texts.

Sicily and the southern kingdom of Italy, which Roger also ruled (to the annoyance of the popes), boasted ports and cities that had retained commerce and communication with the Byzantine Empire—as well as Muslim cities and ports—even after the collapse of the Roman Empire. Their Greek-speaking populations never dwindled as they did elsewhere in Italy and the West, so they were in a better position to raise scholars fluent in Greek and familiar with Arabic and Hebrew.

Chief among the translators of Roger's court was Henricus Aristippus, a contemporary of Gerard of Cremona, working in the West. Unbeknownst to Gerard, Aristippus owned a Greek copy of Ptolemy's *Almagest*, which he had translated—though we don't know by whom—as well as several shorter works of Aristotle. Like Boethius all those centuries before, Aristippus had ambitious plans for translating the major works of Aristotle—*Physics, Metaphysics, On the Soul, On the Heavens*—and like his predecessor, he fell afoul of his lord and was disgraced and dismissed from his role as principal officer of the Sicilian court. He died shortly thereafter in 1162.

Before his death, Aristippus was also responsible for translations of Plato's *Meno* and *Phaedo* and the fourth book of Aristotle's *Meteorology*. He is said to have translated the theological writings of the Christian Father Gregory Nazianzen and Diogenes, though no copies have ever been recovered. Eugene the Emir, fluent in Greek and Arabic, assisted Aristippus and also completed his own translations of Ptolemy's *Optics* (from Arabic), *Catoptrics* of Euclid, *De Motu* of Proclus, and *Pneumatics* of Hero of Alexandria.

The translation movement in Sicily was concerned primarily with ancient works of science, philosophy, medicine, and mathematics. But Eugene also passed on to the West two works of Oriental literature—a prophecy of the Erythrean Sibyl and a Sanskrit fable called *Kalila and Dimna.*

Not long after Gerard's death, Michael the Scot (1175–1232) arrived in Toledo, the one scholar who worked both in Toledo and Sicily. It's not clear whether "the Scot" was originally from Ireland or Scotland (or neither). But he came to Toledo at the close of the twelfth century, and later, between the years 1215 and 1220, he completed an ambitious translation of al-Farabi's massive *De Animalibus* (Book of Animals), a fifteen-volume treatise that included three of the books from Aristotle's earlier treatise on the topic.

This is all the more remarkable in that it was during those years Michael traveled in the company of Toledo's archbishop, Rodrigo Jiménez de Rada, to the Fourth Lateran Council held in Rome in 1215. It was during this time he must have been introduced to King Frederick II Hohenstaufen in Italy, for he dedicated the work to him and for a time served as the Germanic king's court astrologer. Michael returned to Toledo not long after, but in 1229 he left his position at the cathedral for the last time and moved to Sicily for the remainder of his life, translating more books for the charismatic Frederick II.

Of all the royal patrons to support the translation of Greek and Arab science and philosophy into Latin, Frederick II Hohenstaufen (1194–1250) is the most fascinating. In the broad political context of the era, his life seems to have been marked by the failure to absorb Italy and Sicily into the larger Holy Roman Empire. He was excommunicated three times by Pope Gregory IX and appalled many of his fellow princes by the way he (reluctantly) launched the Sixth Crusade, which he brought to a successful albeit brief conclusion by securing the liberty of Jerusalem through a treaty with the sultan of Egypt, al-Kamil.

Frederick made no secret of his admiration for Islam; indeed, it was due to his regard for Arab science as well as his interest in encouraging trade with the Islamic empire that he supported the translation activity at his court. And as we saw in chapter 5, the sultan of Damascus showed his appreciation for Frederick by sending him a great water clock in 1232.

Frederick also loved falconry, another art that was adopted from the Arabs, and this seems to have inspired his interest in broader investigations into nature. After reading a translation of Aristotle's *De animalibus*, Frederick took up the pen himself and wrote *De arte venandi cum avibus* (The Art of Hunting with Birds) in 1248—partly to make known his disappointment with some of Aristotle's scientific observations. "We discovered by hard-won experience," he wrote, "that the deductions of Aristotle, whom we followed when they appealed to our reason, were not entirely to be relied upon, more particularly in his descriptions of the characters of certain birds."[15] Frederick was also disappointed in Aristotle's ignorance of falconry, his adopted favorite sport. He further wrote, "In his work 'Liber Animalium' we find many quotations from other authors whose statements he did not verify and who, in their turn, were not speaking from experience. Entire conviction of the truth never follows mere hearsay."

According one historian, Frederick's statement was revolutionary, "and marks the beginning of the end of the Middle Ages mind-set toward natural history, albeit for the purposes of a sport."[16] It also serves as another reminder of why the translation movement was such a critical factor in the revival of science in medieval Europe. As the cathedral schools grew and evolved into the universities, the masters in the schools all over the continent adopted the newly available texts of the Greeks and the Arabs and incorporated them into a shared curriculum.

The influx of Greek and Arab learning provided European scholars with a standard collection of sources and a common set of intellectual challenges, which, as one historian noted, would prove crucial in the later history of the Scientific Revolution.

> *It was also connected, as both cause and effect, with the high level of mobility of medieval students and professors. Professorial mobility was facilitated by the* ius ubique docendi *(right of teaching anywhere) conferred on the master by virtue of completing his course of study. Thus*

*a scholar who earned his degree at Paris could teach at Oxford without interference and, perhaps more importantly, without acquiring a case of intellectual indigestion; this was possible only because subjects taught at the one did not differ markedly in form or content from those same subjects as taught at the other. For the first time in history, there was an educational effort of international scope, undertaken by scholars conscious of their intellectual and professional unity, offering standardized higher education to generations of students.*[17]

Many have remarked how relatively peaceful was the exchange of texts and ideas among individual scholars of Christians, Jews, and Muslims during the terrible centuries of the First, Second, and Third Crusades. All through this bloody period, the work of translation carried on. By the close of the twelfth century, a large number of the lost works of Aristotle and the great works of the Islamic scientists were making their way, handmade copy by handmade copy, into the schools of France, Italy, Germany to a lesser extent, and England.

It did not take long for the content of the translation movement to have an effect. And Europe would never be the same, however determined some religious authorities were to check its influence. As one scholar noted, "It is a happy irony that one of the great expressions of the cultural unity of the Latin West, evolved at the age of the Crusades, had its roots in the culture which the West was trying to destroy."[18]

The knowledge gained by the translation movement would prove indispensable to the Oxford school of mathematicians who would lead the way to a more quantitative understanding of the physics of motion and whose influence, some historians have argued, would later inspire Descartes, Galileo, and Newton in the Scientific Revolution. This raises the tantalizing question of how things might have turned out had no translation movement ever occurred.

For example, without a recovery of Aristotle's *Physics*, it is difficult to see how the Scientific Revolution would have taken place when it did in Europe, at least in terms of the principles of motion, the three laws ultimately formulated by Isaac Newton in the seventeenth century. Newton's medieval precursors were in their own way reacting to what they perceived to be the faulty principles in Aristotle's physics—particularly in his discussion of projectile motion. They also took issue with his claim that there could be no such thing as a vacuum in nature or that the cosmos is eternal. But the medieval and later Renaissance scholars needed Aristotle's work in detail in order to build off it successfully. And that depended on the translation movement.

# Through a Glass, Not Darkly

*One of the many consequences of the Crusades was a flood of relics: in the Holy Land, the bones of the apostles and the splinters of the Holy Cross were collected by the hundredweight. Those knights and campaigners who returned from the Crusades often enriched themselves with the relics they brought home. From the reconstruction of the skull fragments of Saint John the Baptist, we can assume that this man must have had at least six heads.*

—ROLF WILLACH[1]

As PEOPLE GROW OLDER AND THEIR EYESIGHT BEGINS TO WEAKEN, they expect to need glasses. This is no more remarkable to modern people than the fact that professional athletes tend to peak in their late twenties and early thirties, after which normal wear and tear combined with age leads to the inevitable decline in their performance and often retirement before the age of forty.

As strange as it seems, until the Middle Ages, failing eyesight among the elderly was attributed to disease. Many were the herbal and other remedies that physicians prescribed to cure what they found stubbornly resistant to improving. The invention of reading glasses toward the end of the thirteenth century was, like so many of the inventions we've discussed, the result of a long period of trial and error—and luck. But it showed that nearsightedness or farsightedness could be corrected externally. Over time, the invention also prompted reconsideration of ancient (and erroneous) theories of vision—as well as more detailed theories that were being rediscovered thanks to the translations arriving at the universities.

At the time that spectacles were first invented, glassmakers were as unfamiliar with optical theory as anyone else. This was fortunate. As Vincent Ilardi writes, even if the experienced medieval glassworker could have understood "the formidable intricacies of geometrically based optical theory written or lectured about in Latin, or even explained to him in the vernacular by others, his imagination would have been led in the wrong direction because medieval theory of vision was based on invalid premises."[2] The center of human vision was believed to be at the front of the eye's lens, and rays of light entering the pupil were believed to be refracted on the back side of the lens. Constructing lenses to place in front of the eye made no sense with this theory. But in practice, it worked. It would be centuries before anyone figured out why.

In the meantime, optics remained a theoretical subject of interest only to philosophers and physicists. For example, the great Iraqi astronomer and physicist Ibn al-Haytham (ca. 965 CE–1040 CE) wrote an extensive treatment, *Opticae thesaurus*, as it became known after it was translated and available to the medieval natural philosophers. In al-Haytham's theory, vision was the result of light reaching the eyes in reflection from the objects of attention. His achievement was to demonstrate that such a model was compatible with a geometrical theory of light. And although he employed the mathematics of Euclid to work out precise dynamics of how this worked in a physical sense, he did not break any new ground on the physiology of the human eye itself or how its anatomy might cause lack of visual clarity (nearsightedness or farsightedness).

As Rolf Willach writes, despite several improvements in comparison to ancient Euclid's thoughts on optics, nowhere in the seven sections of al-Haytham's book was there the faintest hint that might have suggested to glassmakers that the idea of grinding pairs of identical convex glass lenses with a refracting power could aid someone with significant farsightedness.[3]

As we now know, in a normal functioning eye,

*parallel rays from distant objects, as well as diverging rays from close objects, will come to proper focus at the retina. If, however, the eyeball*

*is unnaturally elongated, . . . then the rays projected through the crys-*
*talline lens will be brought to focus too early, and the result will be*
*nearsightedness. If the eyeball is unnaturally compressed, . . . the rays*
*will be brought to focus too late, the result being farsightedness. In both*
*cases, the disorder is due to a misshapen eyeball, and the correction*
*entails no more than refocusing the image properly on the retina. That*
*is easily done by interposing a concave or convex lens of appropriate*
*curvature between the visible object and the eye.*[4]

This knowledge made sense only after optics was completely revised, and that did not begin to happen until the time of the great physicist and mathematician Johannes Kepler during the Renaissance.

In the meantime, medieval glassmakers utilized what worked, as the demand for reading aids continued to grow. When the schools associated with the cathedrals began to expand in the twelfth and thirteenth centuries, so did the demand for more books, which meant more work for the monks and clerics copying and proofreading them for the monasteries that produced them. As the older monks and abbots lost the ability to read (and write) clearly, various ancient solutions were tried. For example, using a clear glass globe filled with water as a kind of magnifying glass. One can imagine how tedious and tiring it must have been to roll a glass globe across the page in order to read each line of text, especially in weak candlelight at night.

Such "water globes" date back to antiquity. Indeed, with one example, according to the writings of Pliny the Elder, who was discussed in earlier chapters on the subject of papyrus, plano-convex lenses were used in combination with glass water globes to ignite fires during religious ceremonies in temples or homes and also to cauterize wounds on the battlefield. So the ancients were well aware of the power of magnification. (In volume 10 of his *Natural History*, Pliny sings the praises of glass, writing that no other material known at the time was more pliable or adaptable. The most highly valued, in his view, was transparent, with no color.)

Both convex and concave lenses are based on the idea of a perfect sphere of glass. The plano-convex lens, like the one illustrated on the next page, is "a slice" from the outside of a solid glass sphere, one side of which

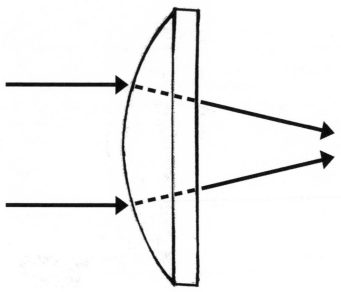

Plano-convex lens. ILLUSTRATION BY RYAN BIRMINGHAM

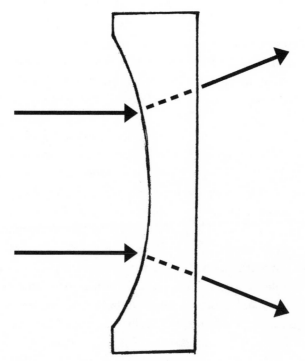

Plano-concave lens. ILLUSTRATION BY RYAN BIRMINGHAM

is ground and polished flat. It creates the effect of magnifying images because the outwardly curved surface refracts—or bends—light rays so that they converge on a single focal point. The concave lens, in contrast, is "a slice" from the inside of a hollow glass sphere. Because it is thicker at the edges than in the center, its inwardly curved surface works to spread light rays apart, making an image appear smaller.

Additional evidence of the existence of both convex and concave lenses during this early period was found in ancient Egyptian statues dating back 4,600 years, which can be found today in the Museum of the Louvre and the Egyptian Museum in Cairo. The glass eyes of the statues were discovered to be ground from high-quality rock crystal. But whatever technique the ancient Egyptians relied on, it was never written down—or if it was, it did not survive and the technique was lost to subsequent centuries. Going back even further, archaeologists have uncovered Bronze Age plano-convex lenses made of ground glass and quartz, many with magnifications of 1.5 or 2 times, making them suitable for reading.

That the craft of lens grinding survived the collapse of the Roman Empire and the turmoil of the barbarian invasions is borne out by more recent findings from Sweden's Gotland Island: ten rock crystal aspherical lenses dating from the eleventh or twelfth century, which appear to have been ground on a turning lathe.[5] Aspherical lenses incorporate some flattening elements to convex lenses in order to improve clarity. Close analysis revealed that the Swedish artifacts approach the quality of modern lenses. Archaeologists believe they were probably looted from their place of manufacture, somewhere in the Byzantine Empire.

If we jump ahead, in particular to the time of the Crusades, when returning knights brought back various relics and many fine examples of polished and ground glass, they inspired (among other things) a fundamental change in the way that churches and cathedrals displayed relics to visiting pilgrims. According to Willach,

*The custom of hiding the relics within a case or cross underwent significant change in the subsequent period. Every town or monastery that acquired one or even several of the relics was proud to exhibit these precious items to all the faithful for their admiration.*

*Therefore the miraculous relics, typically several centimeters in size, began to be presented behind the protection of transparent rock-crystal plates.*[6]

It's interesting that the widespread rise of veneration of relics dates to this period and not earlier in Christianity before the collapse of Rome. But faith in the miraculous effects of ossified body parts, pieces of clothing, or the tools that supposedly belonged to the apostles, Willach explains, was a cult that emerged with the conversion of the barbarians, the Germanic tribes that swept through Europe during Rome's eclipse. It was peculiar to their northern sensibility. "The natural [i.e., pagan] religions didn't understand the admiration of relics," he writes.

Among the treasures brought back from the Crusades were varieties of polished jewels and stones that the monks from the more immediate regions were unfamiliar with. Many of these were incorporated into various relics for display in churches and cathedrals. But it cannot have escaped their notice that many were of such fine quality that they also could be used as reading stones. Their finer quality inspired the monks to devote more resources to grinding and polishing stones of their own, or *lapides ad legendum*, as they called them in Latin.

In particular, they discovered that in addition to their nearly perfectly clear transparency, the rock crystal disks brought back from the Holy Land had convex surfaces of moderate curvature, with the curvature more pronounced at the rims than at the center, which was flatter. As vision aids, what made them dramatically better than the rock crystals they'd been accustomed to using directly on the pages of their parchment books was that these could be held directly in front of the eye to improve vision. Now, instead of sliding a ground rock crystal directly across the page to read, one could hold a more finely ground disk in front of the eyes—in essence, the monks could read with a more convenient vision glass. This is not to be confused with a magnifying glass, which would be more powerful but ineffective too close to the eyes.

Assigning an exact date to this innovation leading to the invention of eyeglasses is not possible, but as Willach writes, we can assume that the optical effect of special moderate curvature of rock crystals was adopted fairly soon after the relics began pouring into Europe after the Crusades during the first half of the twelfth century. "After that date the requirement of such crystal plates for reliquaries steadily increased, and the skilled grinding of such rock crystals became routine. The best of these lenses were separated and mounted in wooden frames for use as vision aids for the older monks."[7]

As much as an improvement as these vision glasses were, they suffered from two defects. One was that the quality of the transparent glass made from rock crystal, which was called white glass, was less than optimal because it tended to be colored around the edges and was replete with small bubbles within the crystal due to chemical residue from the glassmaking process. The second drawback was that no one knew how to precisely grind two lenses of identical curvature for the making of a proper set of eyeglasses, since the grinding technology available did not allow for grinding surfaces with a specific desired curvature. The quality of each white glass rock crystal was different—slightly clearer for one, less clear for another—no two would ever be equal.

The innovations necessary for the first true eyeglasses resulted from two improvements by way of the glassmakers in Venice in the later thirteenth century. One was the discovery of crystal. The second was the cutting of identical lenses from the same single ball of blown glass. These improvements seem to have taken place in rapid succession.

With the end of the Crusades and relative peace among the kingdoms of the eastern Mediterranean, trade began to expand and the Venetians gained an advantage in the glass trade by means of their galley ships. In Egypt, quite possibly at the port city of Alexandria, they discovered natron, a mineral salt from the dried lake beds around Cairo, which the native population had been using since antiquity for various processes—including mummification—going back to the time of the pharaohs. But chief among the mineral's uses was as a component of glass melting. Additionally, the natron from Cairo also contained lime, which acted as a stabilizer in the glassmaking process.

The Venetians found that glass made with natron used as a fluxing agent in the ovens helped reduce the melting point of glass, so that at higher temperatures the glass became more liquid, leaving fewer microbubbles and less color tinge in the finished product once it cooled. The result was glass clearer than ever had been produced before.

The Venetians called this new white glass *cristallum*, and with its clearly superior quality they acted quickly to corner the market on it for the making of glassware. The Venetian guild of eyeglass makers was founded in 1320. By the 1380s, London was importing eyeglasses made from crystal glass at the rate of 384 pairs per month between July and the end of September, according to records, and a hundred years later the rate was 480 per month between November and July.[8]

Somewhere early in this discovery process, a monk from a nearby monastery approached the Venetian glassmakers to suggest they also make single lenses of crystal for vision aid, since they were superior to the white rock crystal the monks were using for their reading aids and much easier to grind. One of the Venetian experts must have questioned the laborious nature of the whole grinding process in the monasteries and suggested that superior vision aids could be made by blowing up a glass ball of a specific diameter for the best magnification and then slicing it into small disks that could be paired, compared, and ground as needed in order to form sets of equal shape and magnification. They could then be mounted in frames in front of the eyes for a more convenient and clearer reading aid than a single magnifying lens.

This was the birth of the first spectacles—or *roida da ogli*, "disks for the eyes," as they were called in Italian. That they were adopted immediately and spread throughout Europe is testified both in letters and in art.

Frames for medieval spectacles. ILLUSTRATION BY RYAN BIRMINGHAM

One example is the sermon delivered by the Dominican Friar Giordano a Rivalto at the church of Santa Maria Novella in Florence on February 23, 1305, in which he expressed his enthusiasm for the new invention: "It is not yet twenty years since the discovery of the art of making eyeglasses for good vision, one of the best and most necessary arts that the world has." Another chronicle from a few years later celebrated a monk at the Dominican monastery of Santa Caterina in Pisa who figured out how to make his own sets of spectacles for the brethren when (presumably) the local inventors were loath to share the trade secrets of how they were made.[9]

By the late 1470s, glassmakers north of the Alps were able to improve on the Venetian technique by cutting a round disk from window glass and then grinding a convex surface for it in a concave mold in Nuremberg. For the purposes of spectacles, they were ground only on the curved side with the planar side untouched. Once Gutenberg's press appeared and the demand for books grew among the public, the demand for spectacles grew as well, inspiring mass production. By the early 1500s, the quality of mold-ground lenses began to decline as the molds themselves were less and less finely measured against a perfect curve. By 1600 the craft of making eyeglasses had declined so precipitously that the first champions of the telescope felt compelled to grind their own lenses rather than use "off the shelf" lenses available from the glassmakers of the time. This strange fact may in turn explain something that has puzzled historians of science—why it took so long to invent the telescope after the invention of eyeglasses in the late thirteenth century.

Not long after, Johannes Lipperhey, a German-born spectacle maker living in Middleburg in the Netherlands, made the first workable telescope in 1608. Within a year, English astronomer Thomas Harriot turned his own telescope to the moon and sketched what he saw. Galileo quickly figured out how to make (or modify) his own telescope the same year, but he ground finer-quality lenses and by January 7, 1610, discovered the moons of Jupiter.

While Galileo utilized his observations primarily for advancing the argument for heliocentrism in his writing and in public argument, others made significant (and until recently, largely overlooked) observations of the planets and stars. Galileo, of course, got into trouble with the Catholic Church for his aggressive championing of Copernicus's theory that the Earth and all the planets revolved around the sun. Although history eventually bore him out, the evidence was not as convincing to his peers (or the church).

Ironically, after being condemned to house arrest for the rest of his life, Galileo's own sight failed, he went completely blind, and no spectacles could help him. He died after his last book was smuggled out of Catholic Italy and printed in Protestant Leyden.

This brings us back to the monasteries of the thirteenth century where the first reading stones and spectacles took off in service of the monks. It's time to examine their role in many of the key inventions of the Middle Ages.

## CHAPTER NINE

# Monks and Nuns Incorporated

*Although a clash was hardly inevitable, it was not long in coming. It seems to have hit hardest at the University of Paris, which not only had the greatest theological school of the Latin Middle Ages but also had one of the best and largest arts faculties. And yet, the conflict that developed must never be allowed to obscure the most important fact: that the translated works of Aristotle were enthusiastically welcomed and highly regarded by both arts masters and theologians. Indeed, Aristotle's philosophy was so warmly received that try as they might, the forces arrayed against it could not prevail.*

—EDWARD GRANT[1]

As WE SAW IN PREVIOUS CHAPTERS, AN IMPORTANT SOURCE OF INSPI-ration for some of the key inventions during the Middle Ages were the bishops and religious orders of the Catholic Church. Particular credit goes to the monks in their monasteries, especially those monasteries run by the Benedictines, whose order dates back to the sixth century CE and who were early adopters of both vertical and horizontal wheeled water mills throughout Europe. A crucial aspect of the Order of St. Benedict, which founded its first monastery in Monte Cassino, Italy, was a dedication to manual labor. In addition to devoting itself to prayer and contemplation, each monastic community was expected to be self-sufficient: to grow its own fruit, vegetables, and grains, to make its own bread, and to raise its own sheep, cattle, and so forth. (To this day, some of the finest liqueurs, ales, and fruit spreads come from European monasteries, and their handcrafted nature offers a distinct selling point over mass-produced products.)

The rise of the monasteries after the fall of the Roman Empire is one of the inspiring stories of the early Middle Ages—at a time when there was not much else to inspire hope. There were monasteries in the Holy Land before the time of St. Benedict, to be sure, but the monastic movement that began in Italy soon accelerated with religious houses being established throughout mainland Europe, Britain, and Ireland. And soon, new orders, such as the Cistercians, the Augustinians, and the Carthusians, took after their Benedictine brothers.

From the beginning, along with their physical labors, the monks also devoted themselves to making careful copies of the scriptures on parchment in order to facilitate the evangelization of the pagan peoples who had invaded the empire. During the bleakest centuries after Rome's collapse, the monasteries also took on the responsibility of retrieving and preserving those texts that had survived. Until the translation movements in the late tenth and eleventh centuries got underway, the monks had no idea how much ancient wisdom, philosophy, and science had been lost in the interim.

As the population in Europe began to grow again in the later eighth and ninth centuries, the monasteries on larger estates invested in more water mills (and later, windmills). They soon found that they were able to provide more than enough food for themselves; they also could supply food for the surrounding community of ordinary folk. This provided an opportunity to sell their surplus for a profit.

There's been a tendency among some scholars in the past to romanticize the role of the monasteries in revitalizing the economies of the Middle Ages and in inspiring innovation.[2] Although credit is due the church for truly revolutionary changes, as we see later, more recent scholarship and evidence has revealed that making a profit and cornering a market when the opportunity arose was as significant an impulse as was an altruistic desire to support the surrounding communities. The larger monastic estates could be as oppressive to the local townsfolk as the most demanding feudal lords.

As Adam Lucas writes, the extent to which a religious establishment was considered virtuous or the opposite, benign or malign, varied greatly according to the size, wealth, and location of the monastery in question.

> *Many religious houses provided alms and legal defense for the poor, board and lodging for the disenfranchised and destitute, care for the sick and aged, and hospitality for travelers and pilgrims, but any examination of an ecclesiastical account book will soon reveal the relatively small sums generally devoted to such activities, even amongst the wealthier houses.*[3]

The monasteries also benefited from the largesse of the secular rulers, who granted them access to large properties for their estates. Under the feudal system of the time, the religious orders and clergy were almost as powerful as the kings and lords of the realm, and they took advantage of their status during the early Middle Ages to dominate the grain milling industry.

> *One of the services that religious houses provided for their own households, servants, retainers and tenants were watermills and windmills to grind their grain. Traditionally, most households had ground their grain at home using a handmill. However, one of the outcomes of the extension of feudal services and obligations throughout medieval Europe between the ninth and thirteenth centuries was the creation of a demesne [manorial] milling sector in which the lower orders were required to grind their grain at their lord's mill. This obligation, known as "suit of mill," generated significant revenues for lords throughout the latter half of the Middle Ages.*[4]

As Lucas shows in the case of England alone, this practice also aroused a great deal of resentment among the common folk. Three major religious orders owned and operated mills throughout England between the eighth and the sixteenth centuries: the Benedictines, the Augustinians, and the Cistercians (with some minor orders thrown into the mix). But by far the Benedictines were the most powerful before the Norman conquest in 1066. "It was they who were the primary benefactors of Anglo-Saxon aristocratic largesse between the early seventh and eleventh centuries and who owned the largest amount of property (including mills) of any of the orders before and after the Conquest."[5]

As the population grew in the tenth and eleventh centuries, along with improved crop yields and commerce, so did the resentment of the ordinary townsfolk and farmers who were compelled to grind their grain only at the mills of the monasteries when they felt it more expedient (and less expensive) to grind their grain at home. Since the monasteries had the backing of the prince or lord of the region for suit of mill, which bestowed on them the exclusive right to charge the peasantry for milling their grain only at their estate mills, they were resented.

In retaliation, the townsfolk often refused to give up their own hand mills at home or went farther afield to the mills of independent manors that would charge less. The abbots in turn sought punishment by appealing to the barons, who sent soldiers to confiscate or destroy private mills and fine the offenders. They often beat them, and in some cases the offender was imprisoned and even executed. There are records of outright riots—for example, the rebellion against the abbots of St. Albans, whose property was besieged by angry townsfolk, as John North writes. (This is the same abbey where Abbot Richard of Wallingford later built his great clock as discussed in chapter 5.)

> *The abbot on this occasion was awarded damages of 100 shillings, an enormous sum, and the man was gaoled—a sentence later exchanged for a fine of 20 shillings. Similar actions were taken against the other townsmen who owed homage to the abbot. From this time on, the same form of protest against the abbot's privileges recurred with increasing frequency. The townsmen beat up one of the monks and damaged a house owned by the abbot in the town. The townsmen of Watford fished in the abbot's fishponds.*[6]

These tensions are not hard to appreciate, as nourishment is the most important need. The most profitable use of the mills during this period was for grains—and this continued to be the case in England, even after the development of mills for other purposes, such as fulling cloth and forging iron, offered mill owners a chance for more profit. But in England, at least, the growth of such mills did not expand dramatically until after the Black Death (1347–1350).

It was on the continent that the real expansion in industrial milling took place. France and Italy took the lead in building mills for driving forge bellows for smithing, crushing bark for tanning leather, crushing ore for metallurgy, and later pounding rags for making paper.

According to Lucas, if we look at industrial mills throughout Europe by type between the years 770 CE and 1600 CE, we can total 4 ore-crushing mills; 6 forge bellows; 8 blast furnaces; 8 olive oil mills; 18 hemp mills for the production of rope; 23 malt mills for the production of ale; 31 tool sharpening mills; 46 sawmills; 58 tanning mills; 236 forge mills; and 635 fulling mills—all powered by water. Though this hardly amounts to an industrial revolution, as some scholars have suggested in the past, it does represent an impressive expansion of economic activity in the later Middle Ages.[7] Although some scholars have argued that the monasteries inspired the expansion of milling into these other industries, by the twelfth century more and more feudal lords and towns were making these investments themselves.

This sense of clerical privilege and ultimate independence from secular authority was also reflected at much higher levels in the church, all the way to the Vatican in what became known as the Investiture Controversy, the great political struggle between the popes and the European kings and princes during the twelfth century. In order to centralize religious and temporal authority in Rome, the popes insisted on the privilege of appointing bishops free of any interference or influence from the secular rulers. Perhaps the most infamous battle in this regard was the martyrdom of St. Thomas Becket, the archbishop of Canterbury, who was murdered by knights on the indirect orders of King Henry II, who had been feuding with Becket on precisely the question of who had jurisdiction in murder cases in England. Becket insisted that accused priests be tried by the church's courts, whereas Henry (not without some justification) insisted that they should be tried in his own courts.

The popes won this battle in the short term, although the church struggled repeatedly with the secular powers for its autonomy. Indeed, in a sense, the struggle continues to this day, as the worldwide clerical abuse scandal of the past few decades has once again raised the question of whether the church can be counted on to govern itself or whether the secular authorities need to intervene. If this political question seems to stray too far from our primary topic of medieval invention, there is a very real sense in which the struggle between popes and kings during this period did indeed lead to some striking innovations that would have repercussions to this day. And that is what might be called the invention of universities, institutions of higher learning that we take for granted, but whose emergence during the Middle Ages might never have occurred but as a "side effect" of the papal legal revolution brought about by the political conflicts between popes and princes mentioned earlier.

The idea of a corporate body or group that could be considered an individual entity by the church and the state—independently of the individual members who made up the body—was not a unique idea prior to the Middle Ages, but its institutionalization by a new system of law certainly was. It is not an exaggeration to say that the emergence of the universities was due to the great legal revolution that took place at the same time. Inspired by the determination of the popes to centralize their own authority, it was also influenced by what was seen as the wider need to organize and codify the different legal traditions that the church and medieval society inherited from the Bible and from the laws of ancient Rome, as well as the folk law traditions of the various Germanic kingdoms that now dominated the continent.

Like so many other inheritances from the ancient world, that of Roman law had been lost for centuries. But as Toby Huff writes, the rediscovery of the Justinian Code of civil law, the *corpus juris civilis*, in the late eleventh century in a manuscript in Italy provided the spark for the development of an entirely new system.

> *The new science of law was the result of simplifying, modifying, and transforming the Roman* corpus juris civilis *by medieval legal scholars. They sought to reconcile the many contrasting and sometimes*

*contradictory elements of Roman law with the Bible, German and other folk laws, as well as Church law. The most important architect of this transformation was a monk named Gratian, who produced what he called a* Harmony of Discordant Canons—*first issued in about 1140—that soon became a canonical text used all across Europe.*[8]

Under Gratian's organizing project, the study of law became an independent science based on three distinct elements, including a body of legal traditions (Roman, Germanic, Judaic), a new method of analyzing the traditions, and an institution where the subject could be studied independently—the universities.

Not to be overlooked during this revolution was the active interest and role of the popes between 1072 and 1122, in particular Pope Gregory VII. As stated earlier, the popes wanted a coherent body of law that would give them the authority to free the Christian church from excessive secular control by the European kings and princes, most especially the interference of kings in appointing bishops and governing the clergy themselves. The most significant result of this revolution "was the declaration of the church's legal autonomy, thereby creating the very idea of separate and autonomous legal jurisdictions between the religious and secular domains."[9]

The result of this papal-led legal revolution was that the Catholic Church effectively became the first modern state. According to Harold Berman, the pope at the crucial time, Gregory VII, "proclaimed the legal supremacy of the pope over all Christians and the legal supremacy of the clergy, under the pope, over all secular authorities. Popes, he said, could depose emperors—and he proceeded to depose Emperor Henry IV. Moreover, Gregory proclaimed that all bishops were to be appointed by the pope and were to be subordinate ultimately to him and not to secular authority."[10]

As noted before, this did not happen without a struggle, and the martyrdom of Thomas Becket stands as the most infamous episode of the beginning of that struggle. Four centuries later, King Henry VIII ended the debate by assuming control of the church in England. Of course the debate had a huge influence on the American Founding Fathers, who took pains to enshrine the separation of church and state in the U.S. Constitution—an article that, ironically, the Catholic Church deplored.

A happy side effect of the church's fight for its autonomy was that this desire for independence soon inspired the nations of Europe to do the same—to form governing bodies independent of their kings and, when necessary, to compel them to abide by the rule of law. Here again, England provides one of the first cases, in which King John was compelled by his barons to sign the Magna Carta, granting them means of legal redress against arbitrary decisions by his majesty.

But back to the universities: at the heart of the legal revolution, Huff notes, was the idea of regarding a collective body of people as a unit, or corporation. Tradesmen already had a tradition of coming together to protect themselves and their interests as a group and had established a right to be represented as a group in the life of their community.[11] It was a small step for the growing communities of scholars at the cathedral schools to organize themselves in the same way—and to put the laws governing their expanding autonomous institutions in writing in a charter granting them the right to appoint or fire teachers, to approve courses of study, and to grant licenses and certificates to the students who passed these courses of study. By a cheerful accident of history, the Latin word *universitas*, which means corporation or "whole body," exclusively came to mean an institution of higher learning.

Guilds and associations were not slow to take advantage of the new opportunities offered by the legal revolution of the Middle Ages, either. Cities and towns also could be incorporated as officially as the schools. Today we take the idea of corporations for granted. Small business owners incorporate themselves, as do nonprofit entities. But the medieval universities were the first great corporations of the Middle Ages—with their own charters, official acknowledgment of their own autonomy.

It was in the universities of the twelfth and thirteenth centuries where the first truly original thinkers of the Middle Ages emerged: Peter Abelard, Robert Grosseteste, Albertus Magnus, Thomas Aquinas, Roger Bacon. They were the products of a new system of education that challenged the religious authorities in a way they had never been challenged before. With the great wealth of science and philosophy texts made available by the translation movement, medieval scholars were faced with defending and rethinking many of the doctrines and beliefs they had inherited since the fall of Rome. Not that these pioneers were not challenged and subjected to accusations of heterodoxy by the religious authorities. But in a sense, the cat was out of the bag once the corporate and legal structure was in place to defend the autonomy of the educational institution. Even bishops scandalized by what was being taught were powerless in the long run to control the universities.

The most famous example from the period was when the archbishop of Paris, Etienne Tempier, tried to have condemned a collection of propositions being taught at the University of Paris. These included some of the writings of the recently deceased Thomas Aquinas, but most of them were the ideas propounded by Aristotle and perhaps his most famous interpreter, the Muslim philosopher and judge Ibn Rushd, known in Europe by his Latin moniker Averroes.

Ibn Rushd endorsed Aristotle's argument that the universe was eternal—it had no beginning in time and would have no end; it had always existed and always would. He further subscribed to Aristotle's belief that the individual soul did not survive the death of the body. As a Muslim, Ibn Rushd had to accept the belief in the immortality of the soul, but he did it by arguing for the existence of a kind of collective soul, a unitary intellect. In this view, the only sense in which immortality could be accepted was in the degree to which the individual's soul includes a reasoning intellective part that is shared by all humans. The idea became known as monopsychism, and it caused a great deal of alarm among Christian theologians teaching at the new universities.

Equally alarming was the notion gaining ground among philosophers that the natural world operated according to its own consistent laws of causality—laws that could be subject to study and exploration by human beings. Although God was accepted as the divine source of creation—the Primary Cause, as he was called by Aquinas—philosophers such as William of Conches, Thierry of Chartres, and Adelard of Bath, who was discussed in chapter 7, argued that the natural world operated according to "secondary" causes that could be studied and explained purely by natural philosophy (science). The concern about this doctrine, which was also accepted by both Albertus Magnus and his most famous pupil, Thomas Aquinas, was that it allowed no room for miracles.

Indeed, even Albert was considered suspect by some theologians for volunteering a natural explanation for the biblical flood in the Book of Genesis. He was dismayed when his prize pupil predeceased him and was accused of promoting heterodox ideas. When Archbishop Tempier's threat to condemn some of Aquinas's writings became widely known, Albert, who would be beatified as the church's patron saint of science one day but was then nearing the age of eighty, undertook an arduous trip to Paris in order to defend Aquinas in no uncertain terms. The story is likely an embellished legend that grew out of the actual events—Albert sending a spirited defense of his pupil in writing. In any event, he succeeded—and just in time. By 1277, Albert and his students were well aware of his growing signs of forgetfulness, and he had to give up his lectures. Although it is not known from what form of dementia he suffered, we do know that by the time of his death in 1280, Albert was in his mid- to late eighties and had retired completely to his monastic cell. When his old friend the archbishop of Cologne came to visit him and knocked on the door, the only reply from within was, "Albert is not here."[12]

Historian David C. Lindberg adds that a particularly important set of the condemned propositions

> dealt with things that God allegedly could not do, because Aristotelian philosophy had demonstrated their impossibility. It was apparently being argued by philosophers that God could not have created additional universes (Aristotle had argued that multiple universes are impossible); that God could not move the outermost heaven of this universe in a straight line (because a vacuum, which Aristotelian philosophy had ruled out, would be left behind in the vacated space); and that God could not create an accidental quality without a subject (for example, redness without something to be red).[13]

Altogether these propositions (among the 219 total) were hastily condemned in 1277 on the grounds that they violated God's divine freedom and omnipotence—that it was wrong to accept or promote any restrictions or constraints on God's power.

What were the effects of these condemnations on the teachers? They were not long lasting. In the first place, according to Lindberg, condemnation of theologically unacceptable theses circulating within the universities was not uncommon.

> *The University of Paris saw at least sixteen such cases in the thirteenth and fourteenth centuries. The condemnation of 1277 differed in its auspices, the range of condemned teachings, and the anonymity of the targeted scholars; whereas the typical censure was internal to the university and directed at specific named masters, the condemnation of 1277 emanated from the local bishop and was aimed at specific ideas or propositions and anybody who held them.*[14]

In the short term, they were a loud reminder to those who taught philosophy at the universities that philosophy was subordinate by God to theology. But the teaching (and the books of Aristotle most cited for attack) could not be banned or stopped completely.

Indeed, some scholars argue that Archbishop Tempier unwittingly did future university philosophers and scientists a favor by inspiring them to seek alternative propositions and theories to those of Aristotle because his were unacceptable for theological reasons. As Lindberg notes, the condemnation had the unforeseen benefit of encouraging the investigation of non-Aristotelian philosophical alternatives, some of which contributed importantly to future scientific developments.[15] Edward Grant also argues that the censure of specific statements led to a rise in hypothetical propositions and hypothetical thinking, which in turn furthered the development of science.[16]

One example targeted what was deemed to be Aristotle's problematic theory of motion—the question of how projectiles once launched could continue in their motion through the air without continuous propulsion by some force from behind. Aristotle tried to make the case that it was the air flowing around and behind a moving object that continued to propel it forward from the rear until friction settled it to the ground.

As far back as the seventh century, Christian philosophers such as John Philoponus found this argument implausible. Indeed, Philoponus went on to outline an alternative theory in which the projectile acquires its motive power, or impetus, from the agent that initially launched it and that this motive power, however temporary, is the cause of its continued motion through the air, until the friction of the air and the force of gravity bring it to a natural state of rest on the ground. Ibn Sina developed this argument further in his *Book of Healing* in 1020 CE, arguing that in a vacuum such an impetus would never dissipate unless some other force acted on the projectile first. By the fourteenth century, Jean Buridan, a teacher at the University of Paris, would propose that the impetus of a projectile was mathematically equal to its weight times its velocity, and later Nicole Oresme further developed this relationship by demonstrating it in the form of graphical illustration (an approach he pioneered).[17] One can see here the groundwork being laid for the mechanics of motion that would later find its fullest mathematical description in the work of Isaac Newton.

In the long term, as noted, the condemnation of 1277 did nothing to rein in the determination of the philosophers to stand their ground and defend their autonomy, which is the key point in our discussion of the universities as autonomous teaching bodies. The philosophers were happy to accept the hierarchical superiority of theology over their discipline (and all others), but they insisted on being allowed to teach and explore it without harassment, which they largely did.

The important point is that the university managed to resist clerical domination, and during the next centuries it evolved into the ideal environment for the rise of science—as well as the Protestant Reformation. This is an institutional factor: as interesting as the individual points of debate are between the bishops and popes concerned about the potentially heterodox opinions of Peter Abelard or Siger of Brabant or even Thomas Aquinas, it was the emergence of this new institution, the autonomous university, that would have a revolutionary impact. The university would become the platform or safe space for scientific speculation and eventually experimentation. If we accept 1543 as the start date for the Scientific Revolution—the year that Copernicus's and Vesalius's books were published—that was still almost three centuries away from the time of the controversies surrounding the condemnation of 1277. Yet already the groundwork was being laid for the institutionalized practice of experimental science, and some scholars at places like Oxford and the University of Paris were already questioning the scientific claims of Aristotle and wondering if the ancient Greeks had been mistaken in some of their assumptions about the natural world.

Perhaps the most striking was the English bishop, Robert Grosseteste, who not only taught himself Greek (and also possibly some Hebrew), but also published his own translations of Aristotle's *On the Heavens* and *Nichomachean Ethics*, along with his commentaries. Grosseteste developed his own "metaphysics of light," arguing that light bestowed form on matter and that the universe itself originated in light before the emergence of matter—an idea no doubt inspired by the Book of Genesis (he was a theologian by training), but also delightfully foreshadowing the modern theory of the Big Bang. Grosseteste has also been credited by some with being the first of the scholastics of the Middle Ages to propose experimentation as a necessary component of the study of natural philosophy, well in advance of the Scientific Revolution, although historians such as John Marenbon have more recently determined that the claim is overstated.

*Although Grosseteste spoke in his commentary on the* Posterior Analytics, *on how to reach a universal principle based on experience (*principium universale experimentale*), the claim someone made that he, first in the Middle Ages, devised an experimental method is borne out neither by his theoretical nor his more practical scientific works. The main achievement of his widely read commentary was, rather, to help put into circulation the ideas of the Aristotelian treatise which did not guide scientific investigation in the modern sense, but helped rather in thinking about the organization of knowledge.*[18]

Another fascinating protoscientist (if you will) of his generation was the English Franciscan Roger Bacon. Although he was not a student of Grosseteste's, he almost certainly knew and was influenced by him and was even more pronounced in his enthusiasm for the systematic study of the natural world than the bishop.

On paper Bacon's career looks impressive, and he might have made more of a name for himself as a precursor of modern science had he not been difficult to interact with personally. He was, as Marenbon writes, the kind of person "that anyone who lives in an academic community even today will have no difficulty in recognizing. Endowed with a brilliant wide-ranging intellect, endless curiosity, limitless energy and unbounded self-confidence, he began to think that only he had a true understanding of the new Aristotelian science and to rail against anyone who did not share his bold and frequently quirky views and plans for educational reform."

The Franciscans tried to silence him, but Bacon published a series of books that he wrote expressly for Pope Clement IV in which he laid out his myriad ideas and observations. Ultimately, Bacon's scientific accomplishments did not live up to his advertisements for them, although he was the first medieval author to systematically investigate what is now known as semiotics—the logic of signs—perhaps because of his talent for multiple languages. His innovations were two: he rejected Aristotle's notion that words signify thoughts, which in turn signify things, arguing instead for a direct connection between words and the things they signify. In addition, Bacon argued that the same word could implicitly signify other things at the same time, a simple example being the word *crane*, which signifies both a type of bird and a device for lifting and moving heavy materials. As familiar as this may seem to people today, no one had thought of this before Bacon.[19]

To be sure, although the necessary institutional developments for the rise of science were put in place by twelfth and thirteenth centuries, the new thinking behind them would take generations to achieve their full effect. Nevertheless, in Huff's view, there is enough evidence to conclude that a vast intellectual and legal revolution occurred in the twelfth and thirteenth centuries in the West, transforming medieval society so that it became a receptive ground for the rise and growth of modern science.

With the growth of the universities came the chance for the sons of even the humblest peasants to acquire an education and the opportunity for something more than tenant farming or a trade. But a question arises: what opportunity did this offer to women in the Middle Ages? The answer is none. Here we are faced with one of the negative side effects of the legal and institutional revolution: women actually lost ground in terms of their ability to join and participate in the reforms that were bringing about the great changes in the church and the new educational institutions.

Because the popes were using their reacquired authority not only to protect the church's autonomy, they were also imposing a long-sought end to the toleration of clerical marriage, which had persisted throughout the first thousand years of Western Christianity. Under Pope Gregory VII and Pope Urban II, celibacy for priests now would be officially mandatory, and the fallout for the existing generation of married priests and their families was brutal, as David F. Noble recounts.

"The clergy were abused and deposed, but their wives were destroyed," he writes. "Abandoned by the church to utter destitution, they and their children confronted the horrors of starvation, prostitution, servitude, murder, and suicide. And if [Pope Gregory's] measures had been harsh with regard to the fate of clerical wives, those of his successor, the Cluniac monk Urban II, were worse."[20] Pope Gregory had been content with ejecting priests and their wives and children from their parish homes and with empowering the secular rulers to enforce the edict on recalcitrants. But Pope Urban, "with a refinement of cruelty," reduced the unfortunate women to slavery and offered their servitude as a bribe to the nobles who should help him in his crackdown on clerical marriage.

Although the nunneries during this era did not receive the same amount of support as the monasteries and nothing like the curriculum being offered in the universities, they did offer an alternative to marriage and childbearing for some women. "For those women not confined to a nunnery," R. N. Swanson tells us,

> home education by tutors, or even a generational transmission from mother to daughter (as later represented in images of St. Anne teaching the Virgin to read) could have conveyed basic literacy, with an emphasis on reading above writing. But this would be to a limited end, to train a good wife. Women were excluded from the main areas

*of intellectual exploration, from the schools, and from the specula-*
*tion of academic theology and philosophy. Being excluded also from*
*the emerging professions—the law, pastoral care, and bureaucratic*
*careers—they are absent from the areas which usually attract histori-*
*ans' attention for the period. This marginalization of women in some*
*ways matches the marginalization of the monastic educational system*
*for men, but with the major difference that monks did integrate them-*
*selves into the new structures in the thirteenth-century universities.*
*Such opportunities were closed to women, even to nuns.*[21]

Judging by written sources alone, the Middle Ages seems like an extended history by and about men, and mostly men of the church, as Robert Fossier writes, "clerics who had no reason to know anything about the body, the head, or the soul of women, which they haughtily ignored."[22] The few women who did write did so from the convent. One of the most influential was Hildegard of Bingen (1098–1179), who published accounts of her visions as well as prophecies, songs, a morality play, and a short handbook on medicine. Most of this she wrote while also serving as the superior of her convent. Another mystic, Elisabeth of Schönau, is known for a vision that might one day be considered prophetic by Catholics no longer persuaded that the clergy should exclude women: that of the Virgin Mary standing by the altar dressed in the vestments of a priest.[23]

The fact remains, however, that women were not a major factor in the twelfth-century transformations. They were affected by them, Swanson writes, especially by the broad social and economic developments as the economies prospered and the population grew. In the cultural sphere they had no direct role or involvement. "Rarely can women be identified as instigators of change; and those whose narratives do stand out remain exceptions to the general pattern. If there was a renaissance for women in the twelfth century, it is still invisible."[24]

What was not invisible was the rise of two parallel and antithetical phenomena: on one hand a strident form of antifeminism asserted in religious and philosophical texts juxtaposed with an idealization of women in the traditions and literature of courtly love, making women an unattainable prize. Both of these developments occurred at a time when medieval Catholicism was being transformed by a surge in devotion to the Virgin Mary.

What women of the time made of all this we do not know, according to Fossier, because they were mute.

> *It is not difficult to discern what they thought if we look to the charges against them. . . . Whatever the context, they thought the contrary and acted accordingly. We can trace female "counterpowers," and I have already touched on them: they appeared around the hearth fire or on the bed pillow; at the "parliament of women" that took place at the fountain, the washing hut, and the mill; at the cemetery, which men feared and avoided; and in the devotions or pilgrimages specific to women. Women were zealous in the cult (or at least the somewhat sulfurous veneration) of the Magdalen, the repentant sinner and "countermark" of the Virgin, saintly or human, while women found a consoling patroness in Mary Magdalen.*[25]

CHAPTER TEN

# Instruments of Discovery

*Admittedly, as an instrument for precise observation the astrolabe was of no great value, while for computation it was usually too small to give more than approximate answers to complex problems. As a teaching device, and for clarifying problems in positional astronomy, it has had few equals.*

—JOHN NORTH[1]

THIS LAST CHAPTER FOCUSES ON THE INVENTION OF INSTRUMENTS that became most crucial to the age of discovery that dawned during the late Middle Ages. Two of these devices could be held in the palm of one's hand: the astrolabe and the magnetic compass. The former helped medieval astronomers put the science of the stars on a more solid mathematical foundation, whereas the latter, in combination with improvements in ship design, helped inspire a new age of trade and exploration. All this occurred even as, in many ways, the medieval world was heading toward some truly catastrophic events.

The astrolabe, or "star taker" as it means in Greek, was once believed to have been invented by the Greek astronomer Hipparchus in Alexandria during the second century BCE. Other legends have attributed it to the genius of Hypatia, the great mathematician of the fourth century. There is no evidence to support this, but the fact that her father Theon wrote a short book on the astrolabe that did not survive antiquity may explain why she was favored.

No astrolabe specimens survive from antiquity. The astrolabe was originally designed to be a portable means of measuring the positions of the sun, moon, and planets against the background of stars based on the user's current latitude. If you didn't know your latitude, you could figure it out by using the alidade, a slide rule mounted on the back of the main disk, to determine the altitude of the North Star (assuming you lived in the northern hemisphere). It also allowed the user to determine the time of day or night based on the stars' positions. For those who could afford one, it's not an exaggeration to say that it represented the first portable computer, however simple in its application.

To an onlooker from a distance, the astrolabe resembled a large clock dial on the end of a chain. Around the rim of the main disk of the device were engraved the 360 degrees of the celestial sphere surrounding the Earth, as well as an inner rim with equal segments marking the twenty-four hours of the day. But the astrolabe worked with more than one dial. The round master disk, or mater, just described, was designed to hold multiple inner plates that corresponded to the user's location; that is, the inner plate was a two-dimensional projection of the local night sky of the user's home latitude—including the meridian, the horizon, and the ecliptic (the path through which the sun travels during the course of the year)—carved in a two-dimensional grid on the plate (usually made of brass). The most familiar of the stars in the "fixed firmament" were carved on very thin branches on a separate cutout plate called the rete, which rested on top of the main disk. By rotating it within the mater, one could line up each star with its correct hourly position on the grid of the celestial sphere visible beneath.

A traveler could have more than one plate. For example, an astronomer from the Holy Land would use a plate designed for a latitude of (roughly) 31 degrees, whereas one at the University of Paris would have one designed for (roughly) 48 degrees.

The earliest astrolabes used by the Greeks were likely very simple, mainly used for determining the altitudes of celestial objects. They might have disappeared from history after the collapse of the Roman Empire had they not been adopted and improved by Arab astronomers starting in the eighth century with the rise of the Islamic empires and their expansion across North Africa and into the Iberian Peninsula.

According to Stephen C. McCluskey, although the device was commonly thought of as an astronomical instrument, in the hands of its Arab adopters, it became much more.

*Astronomical instruments are used principally in scientific research or in teaching; the astrolabe was used more like a watch or a sundial. It was a practical timekeeper that combined a simple device for astronomical observations with scales for astronomical calculations. Yet even as a practical observing instrument it reflects a fundamental change. Heretofore most astronomical observations had been qualitative: observing the appearance or disappearance of a star, noting the changing phases of the Moon, marking the general position of one of the planets in relation to the constellations, or tracing the annual motion of the Sun along the local horizon. On the back of the astrolabe is a sighting vane with which an observer can measure the angular height of the celestial body using a graduated scale on the circumference of the instrument. This measurement is quantitative, yielding a measured angle of a given number of degrees, an angle that fits within the geometrical framework of spherical astronomy.*[2]

The alidade, the sighting vane on the back of the astrolabe, could be used for mundane measurements, such as determining the height of a mountaintop or a tower or the depth of a well. The time of day or night could be determined, roughly, of course, compared to the clocks in our modern cell phones and watches. But as McCluskey emphasizes, the astrolabe was the first instrument to make astronomical observations and predictions more concretely quantitative. This met an immediate need for the people in charge of maintaining official calendars.

As noted in chapter 7, the study of astronomy (and astrology) was hugely important to religious leaders. There was an obvious impetus among them for the adoption and proliferation of astrolabes throughout the Muslim world. As Brian A. Catlos writes in his history of Islamic

Front face of the author's wooden model astrolabe, based on latitude 52 degrees, with removable rete and adjustable measuring vane. PHOTO COURTESY OF THE AUTHOR

Spain, astronomy was a foundational science in al-Andalus, crucial for properly calculating the months and days of the Islamic calendar. This included the beginning and ending of religious feasts, as well as determining what was known as the *qibla*, the true direction of Mecca in Arabia, and the prescribed time for prayers so that prayer could be conducted properly.

*One can see the effect of the study of astronomy on the orientation of mosques in al-Andalus. The Syrians who constructed the great Mosque*

Rear face of the author's wooden model astrolabe, including the alidade for sighting and altitude degrees along the outer rim. PHOTO COURTESY OF THE AUTHOR

*of Córdoba in the eighth century, for example, did not account for the fact that the Earth is spherical and therefore assumed that Mecca was to the south, as it was from Damascus, and thus oriented the mosque in that direction. By the time the mosque of Madinat al-Zahra' was founded, however, the builders were able to better calculate the direction of the Holy City and oriented the prayer hall toward the east. To calculate global position, Islamic thinkers had improved on the ancient astrolabe, an instrument for calculating latitude, and therefore the position of the moon, stars, and planets, for the purposes of reading horoscopes. The instrument would be perfected in al-Andalus with the invention of a more accurate "universal astrolabe," as well as mechanical models of the Ptolemaic universe that better enabled the calculation of planetary latitude and longitude, which made casting horoscopes much more efficient.*[3]

It was through Muslim Spain that the astrolabe first reached Europe in the tenth century. Christian leaders wasted no time in employing it to aid in the determination of their own calendars of religious observances. In 1092, the abbot of the English monastery in Malvern used the astrolabe to work out a table predicting the days of the new moon for several years into the future. Advance knowledge of the dates of the new moon helped determine the yearly date on which Easter Sunday should fall, which in the Western church was always the first Sunday after the new moon following the vernal equinox.

Beyond its use by the church, the astrolabe also became a teaching tool in its own right. The English poet Geoffrey Chaucer, most famous for his collection of pilgrimage stories, *The Canterbury Tales*, which he began composing in 1386, also wrote a *Treatise on the Astrolabe* for a young boy named Louis, who may have been his own son, although it's not certain. Chaucer was a servant of the king at the time—indeed, he served three successive kings, Edward III, the ill-fated Richard II, and his successor Henry IV, about whom Shakespeare later wrote a sequence of his most famous plays. In his professional capacity Chaucer had to travel a great deal, and from one of his visits to the continent he brought back an astrolabe, which he taught himself how to use. The treatise he subsequently wrote was to show young Louis how he could make his very own out of wood.

The astrolabe was also adapted—trimmed down in a sense—to serve as a mariner's tool in the days before the sextant and the arrival of the compass. But this more modest device featured only the mater's trim and the alidade so that sea captains could determine the altitude of a star, the sun, or the moon. It wasn't as useful a navigational tool as the sextant would become.

Although no astrolabe from antiquity has survived, several have survived from the late Middle Ages. The oldest dated European astrolabe is from the fourteenth century and is kept in the British Museum.[4] By the time of the Renaissance, it seems that astrolabes were more coveted as works of art and status symbols, though their educational value never diminished.[5]

Today, a shadow of the astrolabe still exists in the form of cardboard planispheres that one can buy for tracking the stars and constellations. But with the advent of smartphones and apps, even these are giving way to more sophisticated three-dimensional programs that can represent the celestial sphere at a level of detail that no one from the Middle Ages could have imagined.

The astrolabe evolved into something its inventors may never have imagined: the mater, the main disk featuring the hours of the day carefully laid out around the rim, became the prototype for the face of the first mechanical clocks. As we saw in chapter 5, the astrolabe dial was the inspiration for the dials of the great clocks built and mounted in church towers toward the end of the thirteenth century—they were intended, like the astrolabe, to track more than just the hours of the day and the night; in the most ambitious clocks, they were designed to track the motion of the sun, the moon, and even the planets. Even as the astrolabe was first being developed in antiquity, the Chinese were investigating the natural phenomenon that would eventually support a more reliable navigational tool than either the mariner's astrolabe or the sextant. That was the magnet.

The discovery that certain ores were magnetic and attracted metal objects first may have drawn the interest of conjurors and magicians in China, who delighted in persuading patrons that they could move objects about on tables without making physical contact. However, it was not long, historian Amir D. Aczel writes, before higher officials saw a means of exploiting the natural magnetism of the lodestone for more serious purposes.

*An early story known to have been written about 806 BCE describes the palace of Ch'in Shi Huang Ti. The palace had what must have been the first metal-detection system in the world. The entire gate of the palace was made of lodestone, and anyone who tried to enter the palace bearing concealed iron weapons would be detected because of the great magnetic pull of the gate and immediately arrested.*[6]

Experimentation with the naturally magnetized ore revealed that a long slender fragment suspended by a string or floating in a bowl on a piece of wood aligned its ends north and south with the Earth's magnetic field. And further, if you rubbed a piece of iron with a lodestone, its magnetic properties transferred to the iron so that it too became magnetic.

The initial discovery could not help but inspire a certain awe, and Chinese philosophers regarded the lodestone's invisible properties as one of the "laws of heaven." In keeping with the philosophy of feng shui—how to harmonize their living environment according to the energies of wind and water—the Chinese began to orient buildings and temples in the direction of the south, using the lodestone.[7]

But it was not long after, perhaps as early as the first centuries of the Common Era, that the Chinese were sculpting pieces of lodestone into various shapes for use as directional devices that could be cast on the ground to align with the South Pole. According to historian Alan Gurney, the first definite mention of a marine compass in Chinese literature comes from the twelfth-century *Phing-Chou Kho Than* (Phingchow Table-Talk) written by Chu Yü, whose father had been a high official in the port city of Canton.

> *One chapter describes commercial voyages between Canton and Sumatra: "The ship's pilots are acquainted with the configuration of the coasts; at night they steer by the stars, and in the day-time by the sun. In dark weather they look at the south-pointing needle." Chu Yü then continues with a description of the Chinese equivalent of the European lead line: "They also use a line a hundred feet long with a hook at the end, which they let down to take samples of mud from the sea-bottom; by its appearance and smell they can determine their whereabouts."*[8]

Europeans had by this time invented their own version of the compass independently. The first mention of a compass as a navigational tool can be found in the book *De Naturis Rerum* written by Friar Alexander

Neckham in 1187, though he provides no source for how he heard about the device, which, in his words, involved touching a needle to a lodestone and letting it rotate until it came to rest pointing north.

A more concrete description of the navigator's compass was provided later by a soldier named Peter Peregrinus of Maricourt. During the siege of a fortress town of Lucera in southern Italy in 1269, Peter wrote a long letter to a friend in Picardy in which he described two kinds of compass.

One consisted of a carefully cut and shaped lodestone fitted into a circular sealed box with a pointer on the top lid. The box was then to be floated in a slightly larger bowl of water. This contrivance was designed to show astronomers the true meridian without using the sun. The second compass was more sophisticated, consisting of a pivoted, magnetized needle placed in a glass-lidded box. The edge of the lid was graduated into degrees and was provided with sights for taking bearings of stars.[9]

The compass as we know it, however, appears to have attained its final form thanks to the merchants of Amalfi, the prosperous coastal port city south of Naples in Italy, sometime between 1295 and 1302. They designed a round container that protected a mounted magnetic pointer, which could rotate around a compass card that featured eight designated directions: north, northeast, east, southeast, south, southwest, west, and northwest.[10]

Early compass with lodestone floating in a bowl.
ILLUSTRATION BY RYAN BIRMINGHAM

With a compass on board, medieval mariners no longer needed to fear an overcast day (or night), and merchants grew bold enough to ship goods even during the winter months, when the seas and the weather were more dangerous. As seafarers improved their charts, especially in the northern latitudes, they gradually realized that what their compasses were pointing to and what their charts were telling them was true north were not the same thing. This was a new phenomenon they had to account for before drawing up their charts and setting sail.

Magnetic compasses point to the magnetic North Pole, and the angle formed by the difference between magnetic north and the geographic North Pole, called magnetic variation, changes across the world. It also changes over time. "Magnetic variation at London in 1580 was 11°15' East. By 1773, it had swept through 32 degrees to 21°09' West. By 1850, it had increased to 22°24' West. A hundred years later it had decreased to 9°07' West. It is still decreasing today."[11]

The invention of the compass came just as improvements in ship designs were giving merchants the confidence to expand trade and commerce, especially among the northern European kingdoms and the Mediterranean. The evolution of the medieval sailing ship was itself the result of a long process that had begun in the north of Europe with the Vikings before spreading throughout Europe.

Roman galleys and merchant vessels were ideal for the Mediterranean but not for exploration on the open Atlantic. As Richard W. Unger writes, Roman ships were built like fine pieces of furniture. They were strong but expensive and time consuming to build.

*For strength, they relied on the outer shell of the hull. The external planking was placed end to end and held together by mortise and tenon joints. Both the upper and the lower plank were given alternating projections and grooves to fit into each other. Wooden nails, treenails, were then run through the planks and tenons. The treenails in turn were held in place by nails.*[12]

The vessel was finished with a protective covering. "There was no need to caulk the seams—that is, to fill them with some pliable substance—since they were watertight."[13] This approach to ship design, which produced a very durable vessel, dated at least to the time of Homer and appears to have been exclusive to the Greeks and Romans, who designed both their galleys for naval warfare and their merchant vessels using the technique.

The Romans did not harbor ships during their occupation of the British Isles, and after they withdrew, whatever knowledge of shipbuilding they may have imparted to the population disappeared. The design of boats and ships remained quite simple over the following centuries of the early Middle Ages: tub shaped, flat bottomed, and with only a single square sail when they were not powered exclusively by rowers. The canvas-based curragh, still in use by fishermen in some parts of Ireland today, formed a base model for Celtic ship design throughout the early Middle Ages, until the arrival of the Vikings.

Viking ships, although not large, nevertheless served for almost three centuries as a kind of ideal ship, according to Jean Verdon. With its high prow and almost equally high stern raised above the deck and its single mast and single square sail, the Viking ship was usually seventy feet in length and roughly sixteen feet wide. Shields were set along both sides of the ship along the line of holes through which the long oars extended. At maximum capacity, the ship could hold seventy warriors and travel up to 10 knots. Adding to its efficiency was its ability to maneuver, reversing direction within a narrow radius quickly, and to land directly on the shore and to push off quickly.

It took a long time to build such a boat, Verdon writes. Using an ax, Viking shipwrights began by shaping the keel, which served as the central spine of the ship and was usually hewn from a single trunk of oak. Metal rivets or wooden dowels attached it to the stem and the sternpost. The planks were set so that they partially overlapped, in a "clinker" style that was dominant in northern Europe throughout the later Middle Ages.

*These planks were riveted together and the interstices sealed with hemp previously soaked in tar. Varangs, curved or forked pieces carefully placed on the keel along its axis, gave the whole some stability. Transversal beams that kept the varangs apart, longitudinal pieces that crossed the insides of the frames, and the gunwales enclosing the planks completed the ship. The foot of the mast was set on or into the planking. A small platform forward, and possibly another set at the stern, fixed the dimensions of the hold where cargo and horses were stowed.*[14]

A removable prow head, often the sculpted head of a fierce animal or monster, was mounted last of all. Although it must have terrified the residents of the coastal communities the Vikings attacked, its initial purpose on the part of superstitious captains was to ward off the demons and evil spirits of their enemies.

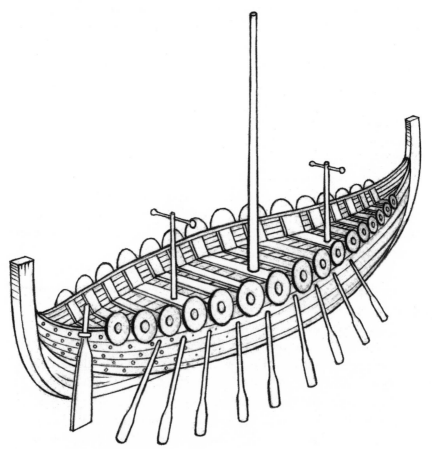

Viking ship design. ILLUSTRATION BY RYAN BIRMINGHAM

Between the ninth and eleventh centuries, the Vikings raided the coastlines of England, Scotland, and Ireland, often founding new settlements, which to this day bear Viking names. But they ventured much farther, reaching as far south as the Iberian Peninsula. (Their Christian descendants, the Normans, later established a kingdom in Sicily.) The same ships also helped the Vikings reach Iceland, Greenland, and even North America, briefly, with the voyage of Leif Erikson. In the long term, the Viking ships were not ideal for transporting goods for significant commerce, however. That required much larger and more durable ships capable of withstanding storms during extended voyages.

A key advance in the design of ships, no doubt driven by the desire to expand trade, took place in the shift from "shell" construction to "skeleton" construction. The former approach entailed building the hull of the ship from the outside in, as the Romans had. By the twelfth century, shipwrights first in northern Europe and then in the Mediterranean began building from the inside out: the keel and the ribs of the ship extending from the keel, making a complete internal frame of the ship before enclosing it in planks. This made for a much more resilient hull. In addition, skeleton-based design allowed builders to complete ships more quickly, and the finished vessels were lighter than those built painstakingly in the older Roman fashion.

In northern Europe, the standard merchant ship initially constructed in shell fashion was the cog, flat bottomed with a single square sail amidships. Once the skeleton-frame approach to design was adopted in the twelfth century, cogs featured a rounder, deeper hull to allow more cargo, as well as a sternpost rudder, an advance in maneuvering that already had been in use by Chinese ships. The sternpost rudder enabled more efficient steering than the classical lateral rudders on either side of the ship. "By the fourteenth century, after being shortened by cutting a port in the stern to put the tiller directly in the helmsman's hand, the strengthened stern-post rudder was also used in the Mediterranean."[15]

Along with changes to the hull of the ship, merchant vessels began adding "castles," or enclosed quarters on the upper deck, both in the front and the rear of the ship, for the use of passengers and for more cargo storage.

The larger the vessel, of course, the more energy needed to drive it, and along with the improvements in the design of the hull, shipwrights also began to add masts and more dynamically adjustable sails. The old square sail of the northern ship was fixed at the top and the bottom and required the entire vessel to be turned to sail before the wind. The new lateen sail, originating in the Mediterranean, resembled a triangle, mounted on the top of the mast via a long crossbar, its front, lower corner tacked down toward the bow of the ship, and its upper rear side reaching back, higher than the mast. The lower corner of the rear of the sail was tied to the stern of the ship, allowing the vessel to tack into the wind. Arab shipbuilders in northern Africa adopted the lateen sail first, and by the eleventh century they had spread to Italian merchant ships.

Another driving force for larger ship design and commerce was the advent of the Crusades. They were important to the Mediterranean cities for both economic and cultural reasons, as Unger writes. The major Crusades of the eleventh and twelfth centuries in particular meant a large and sudden demand for shipping services over long distances. "The involvement of Italian city-states in the fighting drew them both politically and commercially further east. In the end the contest between Christian and Muslim, between the Byzantine and Fatimid Empires, was replaced by competition among the new naval and economic powers in the Mediterranean, Venice, Pisa and Genoa."[16]

Details have survived of the cargo ships that were supplied to King Louis IX of France in his quest to lead the Ninth Crusade in 1271. The king ordered ships from both Genoa and Venice. The largest of the Genoese models carried one hundred horses, along with their knights and attendants.[17]

Notwithstanding the devastating effects of the fourteenth century, the start of which was just a few decades in the future, the people of the Middle Ages had acquired and improved all of the inventions they would need to survive. And survive they would the famines brought on by the increasingly colder climate of the early fourteenth century; the bloody Hundred Years' War between Britain and France; and the onset of the Black Death in 1348, which over the course of three years wiped out almost a quarter of Europe's population. But in less than a century, by the middle of the fifteenth century at the very end of the Middle Ages, European merchants were expanding their operations even more ambitiously. The Portuguese and the Spanish were engaging Italians to build them the ships they would use to begin pushing farther and farther out into the Atlantic.

It's tempting for some historians to view the dark fourteenth century as the end of the Middle Ages, to see Christopher Columbus as belonging instead entirely to the world of the Renaissance ahead of him rather than the long age of building and invention we've been exploring. As historian Peter Denley notes, however revolutionary humanist and Renaissance ideas were to become, their origins were firmly rooted in the culture of the Middle Ages, the age of the clock and the camshaft.

> *In the thirteenth century several features of western European cultural activity were truly international, and Italy participated in these fully along with the rest of Mediterranean Europe. One was the aristocratic culture of chivalric values, courtly love, heroic epic, and so on, much of which had its original focus in southern France, and which is to be found richly in late medieval Spain, Frankish Greece, and to a large extent also in Italy, both in the towns and in the less urbanized south. Another was the deeply religious nature of culture. Religion imbued the Reconquista and the crusading ethos hand in hand with chivalric ideals, and in different forms was prominent in the intellectual ferment of the thirteenth century, the flowering of theology and philosophy with the development of scholasticism, the intellectual stance against heresy, and the organization of preaching. Though Paris was the unquestioned capital of this achievement it was equally an international culture. (St. Thomas Aquinas, the great thirteenth-century theologian of Paris, was an Italian.)*[18]

It's ironic and unfortunate that the scholars who emerged during the Renaissance would be the ones to look back with too much nostalgia to the far-off days of classical Greece and Rome and to dismiss too quickly and ignorantly the more recent age of achievements of their direct ancestors who lived and worked in what they called the "Dark Ages."

# Notes

## Chapter 1: An Ancient Inheritance

1. Maurice Daumas, ed., *A History of Technology & Invention: Progress through the Ages*, vol. 1 (New York: Crown Publishers, 1961), 1.

2. Bryan Ward-Perkins, *The Fall of Rome* (Oxford: Oxford University Press, 2005), 108.

3. M. J. T. Lewis, *Millstone and Hammer: The Origins of Water Power* (Hull, UK: University of Hull Press, 1997), 70.

4. Mark Kurlansky, *Paper: Paging through History* (New York: W. W. Norton, 2016), xiv.

5. Adam R. Lucas, "Technological Change," in *The Encyclopedia of Ancient History*, ed. Roger S. Bagnall, Kai Brodersen, Craige B. Champion, Andrew Erskine, and Sabine R. Huebner (Hoboken, NJ: Wiley Blackwell, 2013), 6559–63.

6. Georges Duby, *Rural Economy and Country Life in the Medieval West* (Philadelphia: University of Pennsylvania Press, 1968), 5.

7. B. H. Slicher van Bath, *The Agrarian History of Western Europe* (London: Edward Arnold, 1963), 78.

8. Georges Duby, *The Early Growth of the European Economy* (Ithaca, NY: Cornell University Press, 1974), 17.

9. Lynn White Jr., *Medieval Technology and Social Change* (London: Oxford University Press, 1962), 43.

10. John Langdon, *Horses, Oxen and Technological Innovation: The Use of Draught Animals in English Farming from 1066–1500* (Cambridge: Cambridge University Press), 10.

11. Jean Gimpel, *The Medieval Machine* (New York: Henry Holt, 1976), 41–43.

12. Adam Lucas, "Narratives of Technological Revolution in the Middle Ages," in *The Handbook of Medieval Studies*, ed. Albrecht Classen (Berlin: Walter de Gruyter, 2010), 19.

## Chapter 2: Harnessing Nature's Power

1. Adam Lucas, *Wind, Water, Work: Ancient and Medieval Milling Technology* (Leiden: Brill, 2011), 52.

2. Örjan Wikander, "The Water Mill," in *Handbook of Ancient Water Technology*, ed. Örjan Wikander (Leiden: Brill, 2000), 375–76.

3. Wikander, "The Water Mill," 395.

4. Wikander, "The Water Mill," 377.

5. Thomas F. Glick, *From Muslim Fortress to Christian Castle: Social and Cultural Change in Medieval Spain* (Manchester, UK: Manchester University Press, 1995), 118–19.

6. See in particular Alexander Jones, *A Portable Cosmos: Revealing the Antikythera Mechanism, Scientific Wonder of the Ancient World* (Oxford: Oxford University Press, 2017).

7. Glick, *From Muslim Fortress to Christian Castle*, 120.

8. Lucas, *Wind, Water, Work*, 38.

9. Lucas, *Wind, Water, Work*, 49.

10. Lucas, *Wind, Water, Work*, 90.

## CHAPTER 3: THE CRANK AND THE CAMSHAFT

1. Örjan Wikander, "Industrial Applications of Water-Power," in *Handbook of Ancient Water Technology*, ed. Örjan Wikander (Leiden: Brill, 2000), 401.

2. Andrew Wilson, "Machines, Power and the Ancient Economy," *The Journal of Roman Studies* 92 (2002): 2.

3. Tullia Ritti, Klaus Grewe, and Paul Kessener, "A Relief of a Water-Powered Stone Saw Mill on a Sarcophagus at Heirapolis and Its Implications," *The Journal of Roman Archeology* 20 (2007): 139–63.

4. Andreas Osiander, *Before the State: Systemic Political Change in the West from the Greeks to the French Revolution* (Oxford: Oxford University Press, 2008), 346–47.

5. Ritti, Grewe, and Kessener, "A Relief of a Water-Powered Stone Saw Mill," 156–57.

6. Terry S. Reynolds, *Stronger Than a Hundred Men: A History of the Vertical Water Wheel*, (Baltimore: Johns Hopkins University Press, 1983), 90–94.

7. M. J. T. Lewis, *Millstone and Hammer: The Origins of Water Power* (Hull, UK: University of Hull Press, 1997), 84.

8. Jean Gimpel, *The Medieval Machine: The Industrial Revolution of the Middle Ages* (New York: Henry Holt, 1976), 13–14.

9. Wikander, "Industrial Applications of Water-Power," 407–9.

10. Reynolds, *Stronger Than a Hundred Men*, 88–89.

11. Lucretius, *On the Nature of Things*, trans. William Ellery Leonard (New York: Dutton, 1957), book V.

12. Norman F. Cantor, *The Civilization of the Middle Ages* (New York: Harper Perennial, 1993), 564–65.

13. Teresa Kwiatkowska, "The Sadness of the Woods in Bright: Deforestation and Conservation in the Middle Ages," *Medievalia* 29 (2007): 40–47.

14. Robert Lacey and Danny Danziger, *The Year 1000: What Life Was Like at the Turn of the First Millennium* (New York: Little, Brown, 1999), 119.

15. Adam Lucas, *Wind, Water, Work: Ancient and Medieval Milling Technology* (Leiden: Brill, 2006), 216.

16. Jeremy Naydler, *In the Shadow of the Machine* (Sussex, UK: Temple Lodge, 2018), 45.

## CHAPTER 4: THE PAPER EXPLOSION

1. Nicholas A. Basbanes, *On Paper* (New York: Knopf Doubleday, 2013), 4, Kindle.
2. Frederick G. Kilgour, *The Evolution of the Book* (Oxford: Oxford University Press, 1998), 12–13.
3. Mark Kurlansky, *Paper: Paging through History* (New York: W. W. Norton, 2016), 9.
4. Pliny the Elder, *Delphi Complete Works of Pliny the Elder* (East Sussex, UK: Delphi Ancient Classics Book, 2015), Kindle.
5. John Romer, ed., *The Egyptian Book of the Dead* (New York: Penguin Classics, 2008), xxxvi.
6. Harry Y. Gamble, *Books and Readers in the Early Church: A History of Early Christian Texts* (New Haven, CT: Yale University Press, 1995), 49.
7. Raymond Clemens and Timothy Graham, *Introduction to Manuscript Studies* (Ithaca, NY: Cornell University Press, 2007), 5.
8. Clemens and Graham, *Introduction to Manuscript Studies*, 9.
9. Kurlansky, *Paper*, xiv.
10. Basbanes, *On Paper*, 6.
11. Kurlansky, *Paper*, 30.
12. Basbanes, *On Paper*, 10.
13. Basbanes, *On Paper*, 10–11.
14. Basbanes, *On Paper*, 58.
15. Kurlansky, *Paper*, 104.
16. Kurlansky, *Paper*, xix.
17. Jeffrey R. Wigelsworth, *Science and Technology in Medieval European Life* (Westport, CT: Greenwood Press, 2006), 69.
18. Wigelsworth, *Science and Technology in Medieval European Life*, 69–70.

## CHAPTER 5: THE GREAT ESCAPEMENT

1. David S. Landes, *Revolution in Time: Clocks and the Making of the Modern World* (Cambridge, MA: Harvard University Press, 1983), 63–64.
2. John North, *God's Clockmaker: Richard Wallingford and the Invention of Time* (London: Continuum, 2005), 164–66.
3. See for example, David S. Landes, *Revolution in Time*. Landes makes a persuasive case that small, simple clocks predated the larger more complex astronomical clocks.
4. Jo Marchant, "In Search of Lost Time," *Nature*, 444 (November 2006): 534–38.
5. North, *God's Clockmaker*, 146.
6. Cited in Gerhard Dohrn-van Rossum, *History of the Hour: Clocks and Modern Temporal Orders*, trans. Thomas Dunlap (Chicago: University of Chicago Press, 1996), 73.
7. Gerhard Dohrn-van Rossum, *History of the Hour: Clocks and Modern Temporal Orders*, trans. Thomas Dunlap (Chicago: University of Chicago Press, 1996), 83.
8. See Lloyd H. Alan, *Some Outstanding Clocks over Seven Hundred Years 1250–1950* (London: Leonard Hill, 1958), 5; and Lynn Thorndike, "Invention of the Mechanical Clock about 1271 A.D." *Speculum* 16, no. 2 (April 1941): 242–43.
9. Lewis Mumford, *Technics and Civilization* (New York: Harcourt Brace, 1934), 17.
10. North, *God's Clockmaker*, 191–92.

## Chapter 6: The Cathedral Crusade

1. Jean Gimpel, *The Medieval Machine: The Industrial Revolution of the Middle Ages* (New York: Henry Holt, 1976), 106.

2. T. Roger Smith, *Architecture: Gothic and Renaissance* (Fairford, UK: E. S. Gorham, 1906), 39.

3. Charles M. Radding and William W. Clark, *Medieval Architecture, Medieval Learning: Builders and Masters in the Age of Romanesque and Gothic* (New Haven, CT: Yale University Press, 1992), 12.

4. Jeffrey R. Wigelsworth, *Science and Technology in Medieval European Life* (Westport, CT: Greenwood Press, 2006), 27.

5. Jonathan Riley-Smith, *The Crusades: A History*, 2nd ed. (New Haven, CT: Yale Nota Bene, 2005), 19.

6. Arnold Pacey, *The Maze of Ingenuity: Ideas and Idealism in the Development of Technology*, 2nd ed. (Cambridge, MA: MIT Press, 1992), 17.

7. Georges Duby, *The Age of the Cathedrals: Art and Society, 980–1420* (Chicago: University of Chicago Press, 1981), 99.

8. R. N. Swanson, *The Twelfth-Century Renaissance* (Manchester, UK: Manchester University Press, 1999), 160.

9. Jon Cannon, *Medieval Church Architecture* (Oxford: Shire Publications, 2014), 27.

10. Wigelsworth, *Science and Technology in Medieval European Life*, 32.

11. Pacey, *The Maze of Ingenuity*, 22.

12. Jean Gimpel, *The Cathedral Builders* (New York: Grove Press, 1983), 58.

13. As quoted in Pacey, *The Maze of Ingenuity*, 25.

14. Robert Mark, "Technological Innovation in High Gothic Architecture," in *Technology and Resource Use in Medieval Europe*, ed. Elizabeth Bradford Smith and Michael Wolfe (Aldershot, UK: Ashgate, 1997), 16.

15. Gimpel, *The Cathedral Builders*, 119.

16. Lorenzo Quilici, "The Study of Roman Roads and Bridges," in *The Oxford Handbook of Engineering and Technology in the Ancient World*, ed. John Peter Oleson (Oxford: Oxford University Press, 2008), 551.

17. Norman F. Cantor, *Inventing the Middle Ages: The Lives, Works, and Ideas of the Great Medievalists of the Twentieth Century* (New York: William Morrow, 1991), 231.

## Chapter 7: From Greek to Arabic and Back Again

1. Marie-Thérèse d'Alverny, "Translations and Translators," in *Renaissance and Renewal in the Twelfth Century*, ed. Robert L. Benson, Giles Constable, and Carol D. Lanham (Toronto: University of Toronto Press, 1991), 459.

2. David C. Lindberg, "The Transmission of Greek and Arabic Learning to the West," in *Science in the Middle Ages* (Chicago: University of Chicago Press, 1978), 58–62.

3. David Levering Lewis, *God's Crucible: Islam and the Making of Europe, 570–1215* (New York: W. W. Norton), 295.

4. David C. Lindberg, *The Beginnings of Western Science: The European Scientific Tradition in Philosophical, Religious, and Institutional Context, Prehistory to A.D. 1450* (Chicago: University of Chicago Press, 2007), 199–203.

5. See Marco Zuccato, "Gerbert of Aurillac and a Tenth-Century Jewish Channel for the Transmission of Arabic Science to the West," *Speculum* 80 (2005): 745.

6. Margaret Gibson, "Adelard of Bath" in *Adelard of Bath: An English Scientist and Arabist of the Early Twelfth Century*, ed. Charles Burnett (London: Warburg Institute, 1987), 13.

7. See, for example, the translation of Adelard in Charles Burnett, *Adelard of Bath, Conversations with His Nephew: "On the Same and the Different," "Questions on Natural Science," and "On Birds"* (Cambridge: Cambridge University Press, 1999).

8. Charles Burnett, *Arabic into Latin in the Middle Ages* (Surrey, UK: Ashgate, 2009), 89–107.

9. Diarmaid MacCullouch, *Christianity: The First Three Thousand Years* (New York: Viking, 2009), 381.

10. Translated by Charles Burnett in his article, "The Coherence of the Arabic-Latin Translation Program in Toledo in the Twelfth Century," *Science in Context* 14, no. 1–2 (2001): 249–88.

11. Charles Homer Haskins, *The Renaissance of the Twelfth Century* (Cambridge, MA: Harvard University Press, 1927), 289.

12. Burnett, "The Coherence of the Arabic-Latin Translation Program in Toledo in the Twelfth Century," 269.

13. Charles Burnett, "Some Comments on the Translating of Works from Arabic into Latin in the Mid-Twelfth Century," in *Orientalische Kultur und Europäisches Mitelalter* (Berlin: Walter de Gruyter, 1985), 166.

14. Thomas F. Glick, *Islamic and Christian Spain in the Early Middle Ages*, 2nd ed. (Leiden: Brill, 2005), 313.

15. Quotes from Frederick, cited by John S. Wilkins, *Species: A History of the Idea* (Berkeley: University of California Press, 2009), are from Casey A. Wood and Florence Marjorie Fyfe, eds., *The Art of Falconry, Being the De arte venandi cum avibus of Frederick II of Hohenstaufen* (Stanford: Stanford University Press, 1943).

16. Wilkins, *Species*, 40.

17. Lindberg, *The Beginnings of Western Science*, 224.

18. Diarmuid MacCulloch, *Christianity: The First Three Thousand Years* (New York: Viking, 2009), 399.


## CHAPTER 8: THROUGH A GLASS, NOT DARKLY

1. Rolf Willach, *The Long Route to the Invention of the Telescope* (Philadelphia: American Philosophical Society, 2008), 17.

2. Vincent Ilardi, *Renaissance Vision from Spectacles to Telescopes* (Philadelphia: American Philosophical Society, 2007), 28.

3. Willach, *The Long Route to the Invention of the Telescope*, 3.

4. A. Mark Smith, "Ptolemy, Alhazen, and Kepler and the Problem of Optical Images," *Arabic Sciences and Philosophy* 8 (1998): 40–42, and cited in Ilardi, *Renaissance Vision from Spectacles to Telescopes*, 29.

5. Ilardi, *Renaissance Vision from Spectacles to Telescopes*, 39.

6. Willach, *The Long Route to the Invention of the Telescope*, 17.

7. Willach, *The Long Route to the Invention of the Telescope*, 24.

8. Frederick G. Kilgour, *The Evolution of the Book* (Oxford: Oxford University Press, 1998), 77.

9. Willach, *The Long Route to the Invention of the Telescope*, 2.

## CHAPTER 9: MONKS AND NUNS INCORPORATED

1. Edward Grant, *The Foundations of Modern Science in the Middle Ages* (Cambridge: Cambridge University Press, 1996), 70.

2. The work of Lynn White Jr. and Georges Duby, for two examples. See White's *Medieval Technology and Social Change* (London: Oxford University Press, 1962) and Duby's *The Early Growth of the European Economy: Warriors and Peasants from the Seventh to the Twelfth Centuries* (Ithaca, NY: Cornell University Press), 1974.

3. Adam Lucas, *Ecclesiastical Lordship, Seigneurial Power and the Commercialization of Milling* (Surrey, UK: Ashgate, 2014), 18.

4. Lucas, *Ecclesiastical Lordship*, 18.

5. Lucas, *Wind, Water, Work: Ancient and Medieval Milling Technology* (Leiden: Brill, 2006), 178.

6. John North, *God's Clockmaker: Richard Wallingford and the Invention of Time* (London: Continuum, 2005), 124.

7. Lucas, Wind, Water, Work, 217.

8. Toby E. Huff, *The Rise of Early Modern Science: Islam, China, and the West*, 3rd ed. (Cambridge: Cambridge University Press, 2017), 112.

9. Huff, *The Rise of Early Modern Science*, 124.

10. Harold J. Berman, *Law and Revolution: The Formation of the Western Legal Tradition* (Cambridge, MA: Harvard University Press, 1983), 94.

11. Huff, *The Rise of Early Modern Science*, 112–13.

12. Simon Tugwell, ed., *Albert and Thomas: Selected Writings* (New York: Paulist Press, 1988), 27.

13. David C. Lindberg, *The Beginnings of Western Science*, 2nd ed. (Chicago: University of Chicago Press, 2007), 247.

14. Lindberg, *The Beginnings of Western Science*, 247.

15. David C. Lindberg, "Science and the Medieval Church," in *The Cambridge History of Science*, volume 2, *Medieval Science*, ed. David C. Lindberg and Michael H. Shank (Cambridge: Cambridge University Press, 2013), 280.

16. See Edward Grant, *The Foundations of Modern Science in the Middle Ages: Their Religious, Institutional, and Intellectual Contexts* (Cambridge: Cambridge University Press, 1996), 70–85.

17. See Lindberg, *Beginnings of Western Science*, 306–313, for a discussion of the evolution of dynamics in medieval thought.

18. John Marenbon, *Medieval Philosophy: An Historical and Philosophical Introduction* (London: Routledge, 2007), 228.

19. Marenbon, *Medieval Philosophy*, 230.

20. David F. Noble, *A World without Women* (Oxford: Oxford University Press, 1993), 133.

21. R. N. Swanson, *The Twelfth-Century Renaissance* (Manchester, UK: Manchester University Press, 1999), 201.

22. Robert Fossier, *The Axe and the Oath: Ordinary Life in the Middle Ages* (Princeton, NJ: Princeton University Press, 2007), 78–79.

23. Elizabeth Alvilda Petroff, ed., *Medieval Women's Visionary Literature* (Oxford: Oxford University Press, 1986), 141.

24. Swanson, *The Twelfth-Century Renaissance*, 202.

25. Fossier, *The Axe and the Oath*, 82.

## CHAPTER 10: INSTRUMENTS OF DISCOVERY

1. John North, *The Norton History of Astronomy and Cosmology* (New York: W. W. Norton, 1995), 131.

2. Stephen C. McCluskey, *Astronomies and Cultures in Early Medieval Europe* (Cambridge: Cambridge University Press, 1998), 173–74.

3. Brian A. Catlos, *Kingdoms of Faith: A New History of Islamic Spain* (New York: Basic Books, 2018), 160–61.

4. Owen Gingerich, *The Eye of Heaven: Ptolemy, Copernicus, Kepler* (New York: American Institute of Physics, 1993), 86.

5. Thony Christie, "The Astrolabe—An Object of Desire," *Renaissance Mathematicus*, April 28, 2016, accessed May 17, 2019, https://thonyc.wordpress.com/2016/04/28/the-astrolabe-an-object-of-desire.

6. Amir D. Aczel, *The Riddle of the Compass* (New York: Harcourt, 2001), 78.

7. Aczel, *The Riddle of the Compass*, 87.

8. Alan Gurney, *Compass: A Story of Exploration and Innovation* (New York: W. W. Norton, 2004), 37.

9. Gurney, *Compass*, 38.

10. Aczel, *The Riddle of the Compass*, 75.

11. Gurney, *Compass*, 27.

12. Richard W. Unger, *The Ship in the Medieval Economy: 600–1600* (London: Croom Helm, 1980), 36.

13. Unger, *The Ship in the Medieval Economy*, 39.

14. Jean Verdon, *Travel in the Middle Ages* (Notre Dame: Notre Dame University Press, 2003), 74.

15. David C. Lindberg and Michael H. Shank, eds., *The Cambridge History of Science*, vol. 2, *Medieval Science* (Cambridge: Cambridge University Press, 2013), 641.

16. Unger, *The Ship in the Medieval Economy*, 120.

17. Unger, *The Ship in the Medieval Economy*, 125.

18. Peter Denley, "The Mediterranean in the Age of the Renaissance," in *The Oxford History of Medieval Europe*, ed. George Holmes (Oxford: Oxford University Press, 1992), 268.

# Bibliography

Aczel, Amir D. *The Riddle of the Compass*. New York: Harcourt, 2001.

Allen, Valerie, and Ruth Evans, eds. *Roadworks: Medieval Britain, Medieval Roads*. Manchester, UK: Manchester University Press, 2016.

Bagnall, Roger, Kai Brodersen, Craige B. Champion, Andrew Erskine, and Sabine R. Huebner, eds. *The Encyclopedia of Ancient History*. Hoboken, NJ: Wiley Blackwell, 2013.

Bartlett, Robert. *The Making of Europe: Conquest, Colonization and Cultural Change 950–1350*. New York: Penguin, 1993.

Basbanes, Nicholas A. *On Paper*. New York: Knopf Doubleday, 2013.

Benson, Robert, Giles Constable, and Carol D. Lanham, eds. *Renaissance and Renewal in the Twelfth Century*. Cambridge, MA: Harvard University Press, 1982.

Berman, Harold J. *Law and Revolution: The Formation of the Western Legal Tradition*. Cambridge, MA: Harvard University Press, 1983.

Bisson, Thomas N. *The Crisis of the Twelfth Century: Power, Lordship, and the Origins of European Government*. Princeton, NJ: Princeton University Press, 2009.

Bloch, Marc. *Land and Work in Medieval Europe: Selected Papers*. New York: Routledge, 1967.

Bradford Smith, Elizabeth, and Michael Wolfe, eds. *Technology and Resource Use in Medieval Europe*. Aldershot, UK: Ashgate, 1997.

Burnett, Charles. *Adelard of Bath, Conversations with His Nephew: "On the Same and the Different," "Questions on Natural Science," and "On Birds."* Cambridge: Cambridge University Press, 1999.

———, ed. *Adelard of Bath: An English Scientist and Arabist of the Early Twelfth Century*. London: Warburg Institute, 1987.

———. *Arabic into Latin in the Middle Ages*. Surrey, UK: Ashgate, 2009.

Cannon, Jon. *Medieval Church Architecture*. Oxford: Shire Publications, 2014.

Cantor, Norman F. *Antiquity: From the Birth of Sumerian Civilization to the Fall of the Roman Empire*. New York: Harper Perennial, 2003.

———. *The Civilization of the Middle Ages*. New York: Harper Perennial, 1993.

———. *Inventing the Middle Ages: The Lives, Works, and Ideas of the Great Medievalists of the Twentieth Century*. New York: William Morrow, 1991.

Catlos, Brian A. *Kingdoms of Faith: A New History of Islamic Spain*. New York: Basic Books, 2018.

Clemens, Raymond, and Timothy Graham. *Introduction to Manuscript Studies*. Ithaca, NY: Cornell University Press, 2007.

Coldstream, Nicola. *Medieval Architecture*. Oxford: Oxford University Press, 2002.

Crombie, A. C. *Robert Grosseteste and the Origins of Experimental Science: 1100–1700*. Oxford: Oxford University Press, 1953.

Daumas, Maurice, ed., *A History of Technology & Invention: Progress through the Ages*. Vol. 1. New York: Crown, 1961.

Dohrn-van Rossum, Gerhard. *History of the Hour: Clocks and Modern Temporal Orders*. Trans. Thomas Dunlap. Chicago: University of Chicago Press, 1996.

Duby, Georges. *The Age of the Cathedrals: Art and Society, 980–1420*. Chicago: University of Chicago Press, 1981.

———. *The Early Growth of the European Economy: Warriors and Peasants from the Seventh to the Twelfth Century*. Ithaca, NY: Cornell University Press, 1974.

———. *Rural Economy and Country Life in the Medieval West*. Philadelphia: University of Pennsylvania Press, 1968.

Eisenstein, Elizabeth L. *The Printing Press as an Agent of Change*. Cambridge: Cambridge University Press, 1979.

Fossier, Robert. *The Axe and the Oath: Ordinary Life in the Middle Ages*. Princeton, NJ: Princeton University Press, 2007.

Freedman, Paul. *Images of the Medieval Peasant*. Stanford, CA: Stanford University Press, 1999.

Freely, John. *Aladdin's Lamp: How Greek Science Came to Europe through the Islamic World*. New York: Vintage Books, 2009.

Fried, Johannes. *The Middle Ages*. Cambridge, MA: Belknap Press, 2015.

Gamble, Harry Y. *Books and Readers in the Early Church: A History of Early Christian Texts*. New Haven, CT: Yale University Press, 1995.

Gies, Frances, and Joseph Gies. *Life in a Medieval Village*. New York: Harper Perennial, 1990.

Gimpel, Jean. *The Cathedral Builders*. New York: Grove Press, 1983.

———. *The Medieval Machine*. New York: Henry Holt, 1976.

Gingerich, Owen. *The Eye of Heaven: Ptolemy, Copernicus, Kepler*. New York: American Institute of Physics, 1993.

Glick, Thomas F. *Islamic and Christian Spain in the Early Middle Ages*. 2nd ed. Leiden: Brill, 2005.

———. *From Muslim Fortress to Christian Castle: Social and Cultural Change in Medieval Spain*. Manchester, UK: Manchester University Press, 1995.

Grant, Edward. *The Foundations of Modern Science in the Middle Ages*. Cambridge: Cambridge University Press, 1996.

———. *Science and Religion 400 BC–AD 1550: From Aristotle to Copernicus*. Baltimore: Johns Hopkins University Press, 2004.

Gurney, Alan. *Compass: A Story of Exploration and Innovation*. New York: W. W. Norton, 2004.

Haskins, Charles Homer. *The Renaissance of the Twelfth Century*. Cambridge, MA: Harvard University Press, 1927.

Hills, Richard L. *Power from Wind: A History of Windmill Technology*. Cambridge: Cambridge University Press, 1994.

Holmes, George, ed. *The Oxford History of Medieval Europe*. Oxford: Oxford University Press, 1992.

Holt, Richard. *The Mills of Medieval England*. Oxford: Basil Blackwell, 1988.

Huff, Toby E. *The Rise of Early Modern Science: Islam, China, and the West*. 2nd ed. Cambridge: Cambridge University Press, 2003.

Hunter, Dard. *Papermaking: The History and Technique of an Ancient Craft*. New York: Dover Publications, 1978.

Hutchinson, Gillian. *Medieval Ships and Shipping*. Rutherford, NJ: Fairleigh Dickinson University Press, 1994.

Ilardi, Vincent. *Renaissance Vision from Spectacles to Telescopes*. Philadelphia: American Philosophical Society, 2007.

Jaki, Stanley L. *Patterns or Principles and Other Essays*. Wilmington, DE: Intercollegiate Studies Institute, 1995.

Jones, Alexander. *A Portable Cosmos: Revealing the Antikythera Mechanism, Scientific Wonder of the Ancient World*. Oxford: Oxford University Press, 2017.

Kealey, Edward J. *Harvesting the Air: Windmill Pioneers in Twelfth-Century England*. Berkeley: University of California Press, 1987.

Kelly, John. *The Great Mortality: An Intimate History of the Black Death, the Most Devastating Plague of All Time*. New York: Harper Perennial, 2005.

Kilgour, Frederick G. *The Evolution of the Book*. Oxford: Oxford University Press, 1998.

Kurlansky, Mark. *Paper: Paging through History*. New York: W. W. Norton, 2016.

Lacey, Robert, and Danny Danziger. *The Year 1000: What Life Was Like at the Turn of the First Millennium*. New York: Little, Brown, 1999.

Landels, J. G. *Engineering in the Ancient World*. Berkeley: University of California Press, 2000.

Landes, David S. *Revolution in Time: Clocks and the Making of the Modern World*. Cambridge, MA: Harvard University Press, 1983.

Langdon, John. *Horses, Oxen and Technological Innovation: The Use of Draught Animals in English Farming from 1066–1500*. Cambridge: Cambridge University Press, 1986.

———. *Mills in the Medieval Economy: England 1300–1540*. Oxford: Oxford University Press, 2004.

Lewis, David Levering. *God's Crucible: Islam and the Making of Europe, 570–1215*. New York: W. W. Norton, 2008.

Lewis, M. J. T. *Millstone and Hammer: The Origins of Water Power*. Hull, UK: University of Hull Press, 1997.

Lindberg, David C. *The Beginnings of Western Science: The European Scientific Tradition in Philosophical, Religious, and Institutional Context, Prehistory to A.D. 1450*. Chicago: University of Chicago Press, 2007.

———, ed. *Science in the Middle Ages*. Chicago: University of Chicago Press, 1978.

———. *Theories of Vision from Al-Kindi to Kepler*. Chicago: University of Chicago Press, 1976.

Lindberg, David C., and Michael H. Shank, eds. *The Cambridge History of Science*. Vol. 2, *Medieval Science*. Cambridge: Cambridge University Press, 2013.

Linehan, Peter, and Janet L. Nelson, eds. *The Medieval World*. London: Routledge, 2001.

Lucas, Adam. *Ecclesiastical Lordship, Seigneurial Power and the Commercialization of Milling*. Surrey, UK: Ashgate, 2014.

———. *Wind, Water, Work: Ancient and Medieval Milling Technology*. Leiden: Brill Academic, 2006.

MacCullouch, Diarmaid. *Christianity: The First Three Thousand Years*. New York: Viking, 2009.

Magnusson, Roberta L. *Water Technology in the Middle Ages: Cities, Monasteries, and Waterworks after the Roman Empire*. Baltimore: Johns Hopkins University Press, 2001.

Marenbon, John. *Medieval Philosophy: An Historical and Philosophical Introduction*. London: Routledge, 2007.

McCluskey, Stephen C. *Astronomies and Cultures in Early Medieval Europe*. Cambridge: Cambridge University Press, 1998.

McLellan, James E., III, and Harold Dorn. *Science and Technology in World History: An Introduction*. 2nd ed. Baltimore: Johns Hopkins University Press, 2006.

Mumford, Lewis. *Technics and Civilization*. New York: Harcourt Brace, 1934.

Naydler, Jeremy. *In the Shadow of the Machine*. Sussex, UK: Temple Lodge, 2018.

Netz, Reviel, and William Noel. *The Archimedes Codex: How a Medieval Prayer Book Is Revealing the True Genius of Antiquity's Greatest Scientist*. New York: De Capo Press, 2007.

Nobel, David F. *A World without Women: The Christian Clerical Culture of Western Science*. New York: Alfred A. Knopf, 1993.

North, John. *God's Clockmaker: Richard Wallingford and the Invention of Time*. London: Continuum, 2005.

Oleson, John Peter, ed. *The Oxford Handbook of Engineering and Technology in the Ancient World*. Oxford: Oxford University Press, 2008.

Pacey, Arnold. *The Maze of Ingenuity: Ideas and Idealism in the Development of Technology*, 2nd ed. Cambridge, MA: MIT Press, 1992.

Petroff, Elizabeth Alvilda, ed. *Medieval Women's Visionary Literature*. Oxford: Oxford University Press, 1986.

Radding, Charles M., and William W. Clark. *Medieval Architecture, Medieval Learning: Builders and Masters in the Age of Romanesque and Gothic*. New Haven, CT: Yale University Press, 1992.

Reynolds, Terry S. *Stronger Than a Hundred Men: A History of the Vertical Water Wheel*. Baltimore: Johns Hopkins University Press, 1983.

Riley-Smith, Jonathan. *The Crusades: A History*. 2nd ed. New Haven, CT: Yale Nota Bene, 2005.

Shippey, Tom. *Laughing Will I Die: Lives and Deaths of the Great Vikings*. London: Reaktion Books, 2018.

Slicher van Bath, B. H. *The Agrarian History of Western Europe*. London: Edward Arnold, 1963.

Smith, A. Mark. *From Sight to Light: The Passage from Ancient to Modern Optics*. Chicago: University of Chicago Press, 2015.

Smith, T. Roger. *Architecture: Gothic and Renaissance*. Fairford, UK: E. S. Gorham, 1906.

Swanson, R. N. *The Twelfth-Century Renaissance*. Manchester, UK: Manchester University Press, 1999.

Tuchman, Barbara W. *A Distant Mirror: The Calamitous 14th Century*. New York: Penguin, 1978.

Tugwell, Simon, ed. *Albert and Thomas: Selected Writings*. New York: Paulist Press, 1988.

Unger, Richard W. *The Ship in the Medieval Economy: 600–1600*. London: Croom Helm, 1980.

Verdon, Jean. *Travel in the Middle Ages*. Notre Dame, IN: Notre Dame University Press, 2003.

Ward-Perkins, Bryan. *The Fall of Rome*. Oxford: Oxford University Press, 2005.

White, Lynn. *Medieval Technology and Social Change*. London: Oxford University Press, 1962.

Whitney, Elspeth. *Medieval Science and Technology*. Westport, CT: Greenwood Press, 2004.

Wickham, Chris. *The Inheritance of Rome: Illuminating the Dark Ages 400–1000*. New York: Penguin, 2009.

Wigelsworth, Jeffrey R. *Science and Technology in Medieval European Life*. Westport, CT: Greenwood Press, 2006.

Wikander, Örjan, ed. *Handbook of Ancient Water Technology*. Leiden: Brill, 2000.

Wilkins, John S. *Species: A History of the Idea*. Berkeley: University of California Press, 2009.

Willach, Rolf. *The Long Route to the Invention of the Telescope*. Philadelphia: American Philosophical Society, 2008.

Williams, Ann, and G. H. Martin, eds. *The Domesday Book: A Complete Translation*. New York: Penguin, 2002.

Yan, Hong-Sen, and Marco Ceccarelli, eds. *International Symposium on History of Machines and Mechanisms: Proceedings of HHM 2008*. Dordrecht, The Netherlands: Springer, 2009.

# Index